Standards, Innovation and Competitiveness

Standards, Innovation and Competitiveness

The Politics and Economics of Standards in Natural and Technical Environments

Edited by R. Hawkins, R. Mansell, J. Skea

Science Policy Research Unit, University of Sussex

Edward Elgar
Aldershot, UK • Brookfield, US

389.6
S785

Published by
Edward Elgar Publishing Limited
Gower House
Croft Road
Aldershot
Hants GU11 3HR
England

Edward Elgar Publishing Company
Old Post Road
Brookfield
Vermont 05036
US

British Library Cataloguing in Publication Data
Standards, Innovation and Competitiveness:
Politics and Economics of Standards in
Natural and Technical Environments
  I. Hawkins, Richard W.
  602.18

Library of Congress Cataloguing in Publication Data
Standards, innovation and competitiveness : the politics and economics
  of standards in natural and technical environments / edited by R.
  Hawkins, R. Mansell, J. Skea.
    p. cm.
  Includes bibliographical references (p. 236) and index.
  1. Standardization—Congresses. I. Hawkins, Richard, 1949–
  II. Mansell, Robin. III. Skea, Jim.
  T59.A1S75 1995
  389'.6—dc20   94–45018
                                                                 CIP

ISBN 1 85898 037 2

Printed and bound in Great Britain by
Hartnolls Limited, Bodmin, Cornwall

Contents

Figures

Tables

Contributors

David Alexander is a private consultant and Technical Director of OSITOP, a consortium of European information technology suppliers and users. He was formerly Strategic Planning Manager in charge of Group Network Services for ICI plc.

Anthony K. Barbour is an independent consultant with many years experience in environmental auditing. He was chair of the pilot study programme on management systems standards for the British Standards Institution.

Stanley M. Besen is a Vice President with Charles River Associates, Washington DC. He was previously a Senior Economist at the RAND Corporation and has taught at Rice University, Columbia University and the Georgetown University Law Center.

Paul A. David is Senior Research Fellow at All Souls College, Oxford and Professor of Economics (and of History, by courtesy) at Stanford University where he was formerly Chair of the Economics Department. He currently holds a visiting Professorship in the Economics of Science and Technology at the Rijksuniversiteit Limburg (Maastricht, Netherlands) and has served recently as a consultant to the OECD and the European Commission.

Dominique Foray is Professor of Economics at the Ecole Centrale in Paris.

Richard W. Hawkins is Deputy Head of the SPRU Centre for Information and Communication Technologies, University of Sussex, and leader of the Centre's research project on telecommunication and IT standardization.

William Leiss is a Fellow of the Royal Society of Canada, and currently holds the Chair of Environmental Studies, School of Public Policy at Queens University, Kingston, Canada. He was formerly Vice-President, Research, and Director of the Centre for Policy Research on Science and Technology at Simon Fraser University, Vancouver, Canada.

François Leveque is Professor and Deputy Director of CERNA, the centre of industrial economics at the Ecole Nationale Supérieure des Mines de Paris.

Richard Lipsey is Professor and Director of the Economic Growth and Policy Program, Canadian Institute for Advanced Research, Simon Fraser University, Vancouver, Canada.

Bengt-Åke Lundvall is Deputy Director of the OECD Directorate for Science, Technology and Industry. He was formerly Professor in the Faculty of Social Science and Economics at the University of Aalborg in Denmark, and has been a member of the Danish Social Science Research Council.

Robin Mansell is Professor and Head of the SPRU Centre for Information and Communication Technologies, University of Sussex.

Kenji Naemura is Professor in the Faculty of Environmental Information, Keio University, Fujisawa, Japan.

Jacques Repussard is Secretary-General of the European Standardization Committee (CEN). He was formerly Assistant to the Director General of the Association Française de Normalisation (AFNOR), and has also been French representative on the European Community Senior Officials Group for Standardization, and the GATT Standards Committee.

Liora Salter holds a joint professorial appointment at Osgoode Law School, and in the Department of Environmental Studies, York University, Toronto.

Jim Skea is Professor and leader of the British Gas/Economic and Social Research Council Programme on Environmental Policy and Regulation at SPRU.

Mary H. Saunders is assistant to the Director of the Office of Standards Services, National Institute of Standards and Technology, Washington DC.

W. Edward Steinmueller is Professor of Economics of Technical Change at the Maastricht Economic Research Institute on Innovation and Technology (MERIT). He was formerly Deputy Director of the Centre for Economic Policy Research at Stanford University.

Gregory Tassey is Senior Economist at the National Institute of Standards and Technology, Gaithersburg, Maryland.

Stanley I. Warshaw is presently Senior Policy Advisor to the International Trade Administration of the Department of Commerce. Formerly, he was Director of the Office of Standards Services at the National Institute of

Standards and Technology in the same Department of the United States government.

Acronyms

ANSI	American National Standards Institute
ANSI X.12	ANSI standard protocol for electronic data interchange
ATM	Asynchronous Transfer Mode
BSI	British Standards Institution
CCITT	International Telecommunication Union, International Consultative Committee on Telegraphy and Telephony
CEN	European Committee for Standardization
CENELEC	European Committee for Electrotechnical Standardization
CFC	Chloro-fluorocarbon
COS	Corporation for Open Systems (US)
DG	Directorate General (of the European Commission)
DSD	Dual System Deutschland
EDI	Electronic Data Interchange
EDIFACT	Electronic Data Interchange for Administration, Commerce and Transport
EMAS	Eco-Management and Auditing Regulation (Europe)
EPHOS	European Procurement Handbook for Open Systems
ESPRIT	European Strategic Programme for Research and Development in Information Technology
ETSI	European Telecommunications Standards Institute
EUREKA	European Research Coordination Agency
FCC	Federal Communications Commission (US)
FDA	Food and Drug Administration (US)
GATT	General Agreement on Tariffs and Trade
GOSIP	Government Open Systems Interconnection Profile
GSC	Global Standards Collaboration Group
HDTV	High Definition Television
HMIP	Her Majesty's Inspectorate of Pollution (UK)
ICT	Information and Communication Technology
IEC	International Electrotechnical Commission
IPC	Integrated Pollution Control (UK)
IPR	Intellectual Property Rights
ISDN	Integrated Services Digital Network
ISO	International Organization for Standardization

ISO 9000	ISO standards series for quality management
ITSC	Interregional Telecommunications Standards Conference
ITU	International Telecommunication Union
JISC	Japanese Industrial Standards Committee
JTC-1	ISO/IEC Joint Technical Committee 1 on Information Technology
LCPD	Large Combustion Plant Directive (European Community)
LEC	Local Exchange Carrier (US)
MAP	Manufacturing Automation Protocol
MITI	Ministry of International Trade and Industry (Japan)
MPT	Ministry of Posts and Telecommunications (Japan)
MS-DOS	Microsoft Disk Operating System
NAFTA	North American Free Trade Agreement
NGO	Non-governmental Organization
NIST	National Institute of Standards and Technology (US)
NRA	National Rivers Authority (UK)
NTSC	National Television Standard Committee, standard for colour television in the US, Canada, Japan and elsewhere
OECD	Organization for Economic Cooperation and Development
OSF	Open Software Foundation (US)
OSI	Open Systems Interconnection Reference Model
OSITOP	European Group on Technical Office Protocols
PAL	Phase Alternating Line, standard for colour television in the UK, Germany and elsewhere
POSI	Promoting Conference for Open Systems Interconnection (Japan)
POTS	Plain Old Telephone Service
PTO	Public Telecommunication Operator
R&D	Research and Development
RBOC	Regional Bell Operating Company (US)
SEA	Single European Act
SECAM	Sequential Couleur à Mémoire, standard for colour television in France, the former Soviet Union and elsewhere
SME	Small or Medium-sized Enterprise
SNA	System Network Architecture
SPAG	Standards Promotion and Applications Group (Europe)
SRR	Social Rate of Return
T-1	US common carrier services providing 1.544 Mbps transmission
T1	American National Standards Institute, Accredited Standards Committee for Telecommunications
TOP	Technical and Office Protocol

TTC	Telecommunications Technology Committee (Japan)
UBA	*Umweltbundesamt*, German Federal Environment Office
UMPD	Uniform Multi-person Prisoner's Dilemma
UNECE	United Nations Economic Commission for Europe
DoD	Department of Defence (US)
VANS	Value Added Network Services
VHS	Standard for video cassettes
WAN	Wide Area Network
X.25	CCITT standard for public packet-switched networks

Acknowledgements

The editors wish to acknowledge the cooperation of the Organization for Economic Cooperation and Development Directorate for Science, Technology and Industry, and especially John Dryden and Georges Ferné, in planning the Workshop in November 1993 which resulted in the papers in this collection. The UK Economic and Social Research Council's Programme on Information and Communication Technologies provided financial support via its contributions to the Science Policy Research Unit Centre for Information and Communication Technologies.

Many people contributed to the planning of the Workshop and the production of this volume. We would particularly like to thank Cynthia Little for her editorial and secretarial assistance, Betty Skolnick for administrative support, and the many postgraduate students who assisted with the organization of the Workshop. Finally, we greatly appreciate the encouragement given by the Director of the Science Policy Research Unit, Professor Michael Gibbons, for an initiative which brought together representatives of the academic, business and government communities to engage in a truly multidisciplinary debate on research and policy issues.

Richard Hawkins
Robin Mansell
Jim Skea
Science Policy Research Unit,
University of Sussex

1. Introduction: addressing the *problématique* of standards and standardization

Richard W. Hawkins

Throughout human history, the invention and application of technology has been accompanied by the development of standards. By the most basic definition, technical standards are agreed external points of reference to which the physical and performance characteristics of technologies can be compared. As such, standards support both conformity and diversity. Standards are obviously inescapable where the production of items of uniform aspect and quality is desired. Less obvious, but no less important for that, standards can also act as a codification of accumulated technological experience – as a 'base-line' from which new technologies emerge.

As phenomena, standards and standardization are at least as complex in their own way as the technologies to which they refer. Indeed, the simple definition given above contains the seeds of at least two profound questions: 'What constitutes an appropriate point of reference?' and 'What constitutes agreement?' In one way or another, all of the papers in this volume are concerned with fundamental questions like these. Indeed, part of the objective of this volume is to present a cross-section of analysis and comment, informed at a high level, in order to help clarify the political, economic, social and even to some extent the technical elements of what we could call the 'standards *problématique*' – the characteristics that set 'standards' apart as a unique area for study by social scientists.

The objective, however, is not simply to define this *problématique* but to suggest ways of addressing it in both intellectual and practical terms. The latter is especially important given that all of the authors represented here are concerned with the identification of directions for future research into standardization issues that bear directly upon the formulation of policy in both the public and private sectors.

The past 20 years have seen a huge proliferation in the number of studies, policy initiatives and press reports on various aspects of standards and standardization processes. A number of strains of theory are now well established

1

in the academic literature, and attention is turning ever more frequently to the planning of empirical studies based upon these theoretical insights. To date, however, studies of standards have been largely sector-specific, thus examining particular standardization exigencies and concerns, and, inevitably, emphasizing certain aspects of the standardization *problématique* over others. Relatively few analysts have attempted to identify generic traits of standardization as an activity, or to compare research data on the form and function of standards across industrial and institutional boundaries.

The papers in this volume were originally presented at the International Workshop on Standards, Innovation, Competitiveness and Policy which took place in November 1993 at the University of Sussex, Brighton, England. The event was organized by the University's Science Policy Research Unit (SPRU) in collaboration with the Directorate for Science, Technology and Industry of the Organization for Economic Cooperation and Development (OECD). The Workshop brought together some 60 senior analysts and practitioners of standardization from 15 countries for three days of intense discussion and debate. The objectives of this Workshop and the potential significance of policy-related research on standards for broader issues of economic, political and social development are outlined by Bengt-Åke Lundvall in Chapter 2.

In an attempt to encourage development of a comparative perspective on standards issues, the Workshop juxtaposed two broad subject areas – Information and Communication Technologies (ICT) and environmental studies. The reason for limiting participation to two areas was partly pragmatic – no single event could hope to accommodate effectively the range of perspectives and opinions that would undoubtedly emerge should too large a sample of technical areas affected by standards be selected. Nevertheless, the selection of ICT and the environment was not random. Not only does standards-making have a particularly high profile in each area, but between them these particular fields arguably have produced the most substantial and mature academic and policy literatures on the subject of standards.

The papers assembled here present a cross-section of ideas and perspectives. Most are written by academics, but a work on standards with the definition of future research agenda as one of its main objectives would be deficient if there were no counterbalancing comment from experienced practitioners and administrators. One of the most difficult issues for research design in a policy context is how to determine the relevance of the research. The Workshop provided a rare and positive opportunity for academic analysts and practitioners to critically examine each others' perceptions of the standards environment. This volume aims to reflect some of the flavour of this useful exchange.

It is of more than passing interest that it is the 'non-technological' disciplines that have spawned most of the academic work on technology standards. Although a significant portion of the training and work of engineers involves the identification of appropriate standards and their application to specific engineering tasks, only a tiny handful of engineering schools worldwide actually treat standards and standards-making as specific subjects. There have been very few analytical (let alone critical) treatments of standards from the technical perspective.

The vast majority of the literature on standards – both technical and nontechnical, academic and industrial – has emerged in the last 20 years, and rigorous empirical studies in the area are only now beginning to accumulate. Why this relatively sudden interest? Part of the answer lies with the re-evaluation in industrial, policy and academic circles alike of the complex characteristics of systems of innovation. The gradual breakdown of confidence in linear models of technology development and diffusion has led to greater analytical interest in inter-firm relationships, collaborative Research and Development (R&D) activity, public/private sector R&D relationships, and so forth as co-determinants in the innovation process. In this setting, standards and standards-making have become more readily identifiable as significant variables in technological development.

Another explanation for the increased interest is that standards issues have become highly visible in numerous 'public policy' and 'public interest' debates. Governments are raising questions about the links between the global standardization system and international competitiveness. Standardization matters are becoming incorporated into public sector industrial policy regimes at an ever increasing rate. Standards questions now fuel public debates concerning the internationalization of production and trading relationships. Standards feature prominently in consumer issues, occupational health and safety concerns, and environmental protection and management debates.

It should not be discounted, however, that one of the principal factors in the current visibility of standards is that certain high-profile industries have characteristics that make them particularly susceptible to the effects of standards. The ICT industries are in this position owing to the growing requirement for interconnectivity and interoperability of systems. However, the fact that studies of standards are most frequently undertaken from the perspective of individual industrial and technological sectors is only partly the result of such practical affinities. The nature of enquiries into standards by social scientists has itself been something of a consequence of the kinds of questions that individual academic disciplines are most prone to ask.

Thus, for political scientists, the standardization *problématique* is most closely related to systems of law and governance. Their primary interest is focused on situations in which standards assume regulatory functions, or where they have implications for the conduct of international relations. On the other hand, for sociologists/historians/philosophers of science and tech-

nology, questions about standards stem largely from concerns about the creation and sustenance of social power structures. From this perspective, the
primary questions tend to focus on institutional processes of standardization,
and on the nature of inputs into those processes. Not surprisingly, standards
questions with an overtly 'social welfare' orientation tend to feature prominently in this line of enquiry.

Economists have tended to concentrate primarily upon the effects that
standards have upon the behaviour of buyers and sellers of technological
products in the marketplace. As Paul David explains below, for economists,
standards questions fall most distinctly into the realm of 'information economics' – the dynamics of market relationships taking account of the amount
of product information available to buyers. A very substantial part of the
economics literature on standards deals with network effects and issues of
technical compatibility. In many ways, the ICT field is virtually a purpose-
built laboratory for economists to explore these relationships and a very large
portion of the economics literature on standards references ICT directly or
indirectly.

It is significant to note, however, that throughout the literature, studies
undertaken from different perspectives have also enticed us with the prospect
that standards and standardization processes might have significant generic
features. In the mid-19th century, for example, Joseph Whitworth (1882), one
of the earliest architects of institutionalized standards-making, observed that
for all of the objective technical orientation, choices in standards-making
were never free of purely subjective criteria (Whitworth, 1882) – an observation that remains at the heart of most studies of standards in our own time.
Similarly, in his pioneering analysis of industrial standardization in the early
US auto industry, the economic historian G.V. Thompson (1954) noted several industrial 'networking' relationships that seem as relevant to today's ICT
industries as they were to systems of component production early in the
century.

Despite obvious affinities for certain subject areas, economists have not
restricted themselves to discussions of compatibility standards in ICT any
more than sociologists, political scientists and historians have restricted themselves to highly politicized 'social' issues. As these papers will demonstrate,
many common concerns permeate the whole corpus of social science research on standards, and theory developed with particular reference to one
industrial situation is frequently visited in the context of another.

In examining the role of standards in environmental policy, political phi-
losophers can be just as concerned with the nature and function of 'information' and the phenomena of 'network' relationships as economists are concerned with the costs of adoption or non-adoption of computer interface
standards. Moreover, decisions on the environment are also made for eco-

nomic reasons. Thus, in this volume, François Leveque applies similar economically oriented standards-selection criteria to potentially 'life-critical' environmental situations as Dominique Foray and Edward Steinmueller apply to 'commercially critical' situations related to ICT.

Likewise, many concerns are shared not only across academic disciplines, but also across the gap that often exists between the academic analysts and the standards practitioners and policy makers. The work presented here is concerned with such issues as supplier/user relationships in standards-making, determination and application of rules governing participation and decision-making in standards organizations, the public and private sector agenda in seeking voluntary standards, and the ultimate implications of standards for innovation and competitiveness. For administrators and standards practitioners, this is the stuff of which everyday problems are made. For academics, these issues present intellectual challenges of potentially great practical relevance.

Thus, from a non-academic perspective, Stanley Warshaw and Jacques Repussard (respectively senior administrators from US and European standards bodies) present these issues in terms of the institutional positions fostered by fundamental historical discrepancies in political and cultural attitudes towards the function of standards and the role of the public sector in determining and applying them. On the ICT side, taking account of the current business environment, David Alexander places the issues firmly in the context of the pragmatics of standards for technology application in user firms. On the environmental side, Anthony Barbour examines the role of standards in environmental audits, and their potential role in linking corporate commercial decision-making to increasing public expectations regarding environmental protection and management.

From an academic and policy-oriented point of view, Robin Mansell discusses the position of standards-making with respect to the concept of 'institutionality' in political and economic relationships. Liora Salter and William Leiss frame the issues in the social and cultural contexts of building policy communities and dealing with the dynamics of 'stakeholder' interests. Stanley Besen, François Leveque, Jim Skea, and Richard Hawkins examine some of the historical and theoretical foundations of institutional practices in setting standards. Kenji Naemura and Dominique Foray both examine the role of technology users in setting technical standards – the former offering an account of the problems and historical institutional responses, and the latter dealing with the theoretical issues. Gregory Tassey considers the role of standards in establishing technological infrastructures to support future innovation. Edward Steinmueller offers a theoretical analysis of the interplay between political and commercial exigencies in determining the basis for standardization. Paul David critically examines standards in relation to the 'freedom versus order' debate, and, in the process, explores the nature of standards as an issue in economic thinking and as a problem for policy makers.

As most of the chapters attest, the standards enterprise, long characterized by an image of cooperation and harmony, has entered a time of division and dissension. Reflecting the current political and economic climates, as well as the ever increasing rapidity of technical change, the standardization community is now in an unprecedented state of tension. The entire institutional structure of standards-making is under extreme scrutiny by industry and government alike, and standards development organizations must now look for new explanations and justifications for their activities.

As a consequence, this volume does not present the issues and problems, let alone the proposed options and solutions, as part of some artificially harmonized panorama. The authors do not always agree and no editorial 'magic wand' has been employed to make it appear as if they do. Where standards are concerned, no overview of the present state of affairs would be accurate if it did not reflect both the diversity of opinion amongst academics, policy makers and industrialists, and the differences that often exist in the ways academics and practitioners perceive the standards environment.

2.   Standards in an innovative world

Bengt-Åke Lundvall

The various actors involved in the daily business of standardization on the broad front of industrial activities have never gone out of their way to make it easy for laypeople to comprehend what this business was about. Oceans of acronyms and coded language have always proven effective barriers to outsiders, and standards-related discussions have seemed forever doomed to be conducted by faceless technicians in secluded locations.

Yet policy makers and scholars express more and more often their conviction that standards are important. A very small number have actually entered these unfamiliar grounds. The results have been astounding. On the research side, historians, economists and sociologists have produced ample evidence that standards have played, and will play even more so in the future, a key role at the core of technological processes, innovation and industrial development. On the policy side, standards have more and more frequently emerged at the centre of fierce political debates at national and international levels.

Since the middle of the last decade, the Organization for Economic Cooperation and Development (OECD), a policy-oriented international government organization, has taken a growing interest in standardization questions, particularly to bring to light the economic implications for its Information, Computer and Communications Policy Committee. This may truly be a sign of the times. It is no accident that standards are increasingly subject to curiosity or caught in unexpected zones of political turbulence: their place is being redefined, in line with that of technology in our economies and societies and as a consequence of the fact that they are bound to become a major infrastructural component in the development of a global economy.

In this Chapter, I shall briefly recount the various factors that are gradually pushing standards to the forefront of the technology policy scene, before outlining why, in my view, the two domains of Information and Communication Technology (ICT) and the environment have been singled out as foci. Finally, I shall comment briefly on the research and policy agenda.

THE NEW STRATEGIC IMPORTANCE OF STANDARDS

In one form or another standards are everywhere, in the realm of nature, human artefacts and also human thought. They have long been ignored in our attempts to understand technological change, but we realize today that much of the work of historians or economists of technology has actually dealt with standards rather than with technology as such – in a way, it is as if they had thought to examine the forest but really looked at the trees. The process of development and accumulation of technical standards is thus as old as humanity and it is because technology today has acquired greater importance and reached levels of unprecedented complexity and universality that this inheritance may in fact constitute a problem. In technology, the old has always co-existed with the new and one of the driving forces behind the generation of standards has always been the need to make possible this type of technological blending.

The question today is whether the old processes and approaches are still relevant and can continue to be as effective as in the past. Three series of major changes in particular have taken place which directly affect the implications of standardization. Firstly, the stakes have become extremely high as the new ICTs provide opportunities for the establishment of a world-wide universal infrastructure needed by the emerging global information economy. Secondly, the conflicts of interest have become correspondingly greater, and now involve regions and nations as well as very large multinational corporations whose future may depend on the selection of a particular standard. Thirdly, a new urgency is felt because of the world *problématique* of global threats to the environment that may call for more rapid and more international development of new understandings and decisions with regard to the choice and implementation of standards.

This is ample evidence that major adjustments are needed to meet demands in the standardization process that have increased as a result of technological and economic change. Yet, established organizations and procedures are not necessarily geared to such an acceleration of their operations. In fact, recent political and social events – from deregulation of key sectors such as telecommunication to a new awareness on the part of the general public and the users of certain technologies, and including the emergence of new standardization fora at international level – have increased the number and types of actors who can claim to be recognized as 'stakeholders' in the standardization process and, hence, may legitimately expect to be involved in it. One consequence has been the multiplication of *de facto* standards and related proprietary claims, going against the prevailing feeling that an effort is needed to streamline mechanisms, bring greater coherence, foster easier interfaces, accelerate the work and pay greater attention to the various non-technical implications of choices being considered.

For example, the potential impacts of prospective standards on the economic, environmental, cultural, social and political spheres may need to be taken into account to a larger extent than has been the case in the framework of highly specialized technical organizations. Consideration of the financial consequences of selecting a particular standard in terms of migration of users, or of the costs of, for example, conformance testing requirements, should be part of a standardization system.

For a number of reasons, all of these new preoccupations may in fact suggest that more active government attention and intervention may be needed. Firstly, the increasing complexity of the standardization work itself will entail higher costs. It is not clear how these can be met when the work has traditionally been voluntary and government support limited in many countries. Secondly, it has become apparent that market forces alone do not necessarily ensure that the best standard – in economic as well as in social terms – will necessarily prevail. Thirdly, the general trend in democratic societies fosters greater transparency in all decision-making processes while traditional standardization discussions aim at achieving consensus and have been better served by less explicit identification of the various interests involved. In the future, and given the foreseeable extension of the range of private interests at stake, governments may have to be more directly involved as representatives of the public interest.

These are but a few of the considerations to be taken into account, but they suffice to illustrate the fact that, although they are all vital, they are often contradictory. The involvement of users, for example, is obviously very important for a number of reasons – not least because users have a major role to play in the innovation process and the development of better technologies; I have attempted to demonstrate this in my own past research into product innovation in the context of user–producer interaction (Lundvall, 1988; 1992). However, the involvement of users will obviously increase the cost and time needed to produce standards and the participation of various groups representing heterogeneous interests will add to the problem of achieving consensus.

Questions such as these are now being debated in the standardization and government communities. Actions will not be effective, however, if we do not attain a better understanding of the basic processes at work. The focus on ICT and the environment singles out two areas in which the implications of these issues can be discussed with regard to different technology, policy and institutional contexts.

STANDARDS IN TWO DIFFERENT SETTINGS

At first glance, the standardization requirements of ICT and the environment seem to be drastically different. On the one hand, the ICT sphere can be characterized by its high rate of technological change and the fact that it is increasingly viewed as a single system – a network for the gathering, transportation and processing of information – flexible enough to be applied in each case according to the specific needs of each user. The successful deployment of these network technologies, at national and international level, will require basic interface standards compatible with requirements of communication and portability. However, the broad menu of possible applications in this technological area is so flexible that it cannot be fully predetermined and is thus not very conducive to the production of detailed standards. The trend will be towards broad functional standards that can be taken into account more easily in diverse contexts.

Environmental standards, on the other hand, most often relate to production methods and/or products which are to meet certain specifications. They may be formulated in terms of absolute targets or thresholds to be observed ('zero pollution' or 'not more than x per cent of a given substance') or in terms of qualitative specifications to be observed ('the material should have the following characteristics ...'). In all of these cases, specific environmental concerns will lead to fairly precise standards.

The products of the standardization work in the ICT and environment fields are thus very different, the former tilting towards interface and functional standards, and the latter towards dimensional and quality standards. However, in spite of these differences, the two standardization areas have important common features and issues.

The question of user participation, for example, has become a central concern in both cases. In environmental standardization, the problem is essentially to ensure fair representation of an undefined public interest in the definition of standards that may affect overall or particular living and working conditions. In ICT, one seeks to involve users as early as possible in the formulation of standards that might thus better reflect their specific needs and conditions. In both cases, however, it is the very legitimacy of standardization methods that is at stake, and this question raises further questions regarding the cost and institutional consequences of expanding involvement of users, the transparency and procedures followed in various fora, and so on. These user-related concerns may thus have a similar impact on standardization processes.

In both areas, the need to adapt to the rapid rates of technological change, and to encourage standardization discussion at an early stage in the development of new technologies, has prompted an interest in new approaches to standardization. The quest for consensus will be facilitated when it is sought at a precompetitive stage, if the various firms have not yet embarked upon open competition in their different markets. Anticipatory standards or even broad functional standards may thus offer attractive alternatives in both areas to the traditional patterns of discussion following the diffusion of new products or processes. Special procedures are certainly needed in such cases, as well as better understanding of the resulting impacts on future technological developments. The fact that emerging technologies do not have users that can be consulted complicates these matters still further.

Beyond these two limited illustrations of a certain convergence of concerns, there is of course the obvious fact that we are dealing with the two technological fields that have had, and will undoubtedly continue to have, a great impact on our economies and societies. These two endeavours are no longer as contradictory as they might have seemed: ICT now holds the promise of increasingly effective environmental monitoring and prevention of environmental damages through the more general application of sophisticated information management systems. Such goals can only be achieved through the application of appropriate standards that will have a long-term beneficial impact on the technological trajectories of the future.

TOWARDS A RESEARCH AND POLICY AGENDA

Standardization has become an enormous world-wide business. It is only a reflection of many years of neglect that we know so little about the amount of resources devoted to this work by the public and the private sectors. It may well be that there are more engineers working on standards today than are directly involved in innovation-oriented R&D projects. Yet standards are also at the core of the innovation process in modern societies in many ways that are not yet fully understood. Better insights into these processes would certainly improve our general understanding of the factors that facilitate or impede technological innovation. At the same time, such insights would make it easier to design standards taking into account these potential impacts.

It is clear that the future requirements of the emerging global economy will become even more pressing in their demands for more and more complex interfaces, greater integration of systems and removal of the various obstacles (some of which might even be technical standards) that hamper international trade or promote unfair competition.

The present focus is on attempting to create comparative perspectives on the study of standards in two major areas of research. However, one should not overlook the growing convergence of the two areas. In a sense, the environment becomes a 'user' of information technology. More generally speaking, the merging of two technical cultures towards the production of effective interfaces between the techno-mechanical and the biotechnical spheres is at stake. These concerns already represent major challenges for the standardization system. Standards have become a pervasive, cross-sectoral and transnational phenomenon and are directly relevant to the design and governability of our complex technical system. It is in this light that the research agenda, which I hope will emerge, will be of the highest relevance for policy makers.

Considering the challenge we have to meet, it is worth recalling what I stressed earlier: standards have long been neglected because they were thought to have limited intellectual or policy interest. We now must learn that the efficient management of the standardization system will provide an essential foundation for future prosperity, growth and employment in our countries. The ways in which we manage to cope with these requirements will above all have a considerable impact on the functioning of the multilateral system of international relations.

PART 1

The public and private interest in standardization

L 96
038

3. Standardization policies for network technologies: the flux between freedom and order revisited

Paul A. David

THE ARCANE WORLD OF TECHNICAL STANDARDS

Once upon a time, in a simpler world, the business of setting technical standards was not an item on the agenda of economists and political scientists. It was held to be one of those arcane and tedious matters best relegated to the attention of engineers. During the past decade, however, standards and standards-setting have emerged as subjects of strategic economic importance demanding the attention of corporate executives and research managers, especially those whose firms are in the business of supplying equipment, operating networks and providing enhanced network services in the computer and telecommunication industries. The existence and nature of technical standards are of vital concern for the encoding, storage processing and transmission of data. Questions concerning the compatibility of the components of information systems and the roles played by market competition, voluntary standards-setting organizations and government regulators in creating standards that affect compatibility, have therefore emerged as critically important in the development and marketing of a wide range of information technology products such as computer operating systems and software, value added data networks, local area networks, high definition television and the like. This is hardly surprising, in view of the fact that technical compatibility issues are of such obvious importance in these areas.

Concomitantly, the volume of activity being carried on by voluntary and *quasi*-governmental organizations charged with the work of developing and publishing such standards also has been greatly expanded, as is reflected in the rapid growth in the numbers of technical committees and working groups, draft standards, and published 'recommendations' at both the national and international levels (US Congress, 1992; Noam, 1992). Although voluntary standards organizations have been with us for a long time even in the international arena, and their activities cover a wide array of conventional industries, their importance has grown in recent years largely due to the developments in micro-electronics and telecommunication technologies.

In the wake of all these developments it is gratifying that a growing number of economists, particularly those who have concerned themselves with technological change and industrial organization problems, have begun to make some headway towards understanding why the setting of standards has not been left to the unfettered workings of markets in the past, why public policy regarding standardization should be a matter of concern not only to the specialists in the industries involved, and why the existing institutional and organizational structures that once seemed adequate to attend to the matter should now appear so beleaguered by critics and targeted for reforms that run the gamut from proposals for privatization to the imposition of stronger governmental regulation and control. In this Chapter, I shall try to step back a bit from the details of contemporary discussions and the myriad debates about specific public policy problems. I shall try to set out what I believe economic analysis has been able to tell us about the way best to think about standards and standards development institutions in network industries and to contribute thereby to distinguishing policy approaches that might be classed as 'generic' to all standardization activities from ones that are appropriate only to particular industrial contexts.

A few terminological preliminaries are in order inasmuch as discussions of standards and standards institutions are made more obscure and confusing by the use of language whose meaning is not context-independent: the term 'standard' itself is used to refer to a document, which can be conceptualized as an information product, and, in other contexts, to refer to the technical specifications or operating characteristics of tangible, physical commodities of varying degrees of complexity. Generically, a 'standard' is to be understood, for the present purposes, as a set of technical specifications that may be adhered to by a producer, either tacitly or as a result of a formal agreement. It is helpful to distinguish among several kinds of standards – reference, minimum quality and interface or compatibility standards (David, 1987). In the telecommunication and computer technology fields, certification of conformity to reference and minimum quality standards provides notice to the user that a given product or component can be successfully incorporated in a larger system comprised of closely specified inputs and outputs. A product that conforms to an interface standard can serve as a subsystem within a larger system built from numerous components and subsystems that are provided by different suppliers, each of which also conforms to the same standard.

Just as there are many different types of standards they can also arise in a number of ways. 'Voluntary standards' is a term used in reference to sets of technical specifications formulated with this ostensible purpose in view; documents of that kind have been described as 'agreements intended to facilitate communication within an industry' (David, 1987: p. 212). Following the taxonomy proposed by David and Greenstein (1990), we may distinguish among: (a) unsponsored standards, these being sets of specifications that have no identified originator holding a proprietary interest, nor any subsequent sponsoring agency, but nevertheless exist in a well-documented form in the public domain; (b) sponsored standards, where one or more sponsoring entities holding a direct or indirect proprietary interest – suppliers or users, and private cooperative ventures into which firms may enter – create inducements for other firms to adopt particular sets of technical specifications; (c) standards agreements arrived at within, and published by, voluntary standards-writing institutions such as the organizations belonging to the American National Standards Institute (ANSI); and (d) mandated standards, which are promulgated by governmental agencies that have some regulatory authority.

The first two of the foregoing outcomes emerge from market-mediated processes and are referred to generally as *de facto* standards. The latter pair usually issue from political ('committee') deliberations or administrative procedures which may be influenced by market processes without reflecting them in any simple way; both are sometimes tagged loosely as *de jure* standards, although when the standards actually do have the force of law behind them, as in the lattermost case (d), it is convenient to refer to them as technical 'regulations'. Although the International Telecommunication Union (ITU) is a treaty organization and its regulations are binding upon the national signatories, the ITU's Telecommunication Standardization Bureau (formerly the CCITT) and other international bodies are only able to issue technical specifications in the form of 'recommendations'. The focus of the present discussion is upon the non-regulatory (*de facto* and *de jure*) processes through which national and international technical standards arise and will touch on the need to consider the interactions between standards-setting effected through competition in markets and through debates and negotiations in committee rooms. However, it should be evident that all the sub-processes of this 'system' are operating concurrently, and, as is clear in the case of the US Federal Communications Commission's (FCC) deliberations over a standard for digital High Definition Television (HDTV), regulatory standards-setting may have direct and powerful impacts upon the other sub-processes, as well as more generally interacting with them in complex ways (Cave and Shurmer, 1991).

CHOOSING BETWEEN FREEDOM AND ORDER: A METAPHOR FOR THE STANDARDS POLICY PROBLEM?

The kernel of the problem posed for private and public decision-making with regard to the setting of technology standards may be construed to be nothing more and nothing less than the fundamental issue with which all social organizations are confronted: where to position themselves on the terrain between the poles of 'order' and 'freedom'. One might well say that the landscape of human social arrangements in this regard is *fractal*, for, at whatever level of detail we choose to observe it, there we can find tensions and organizational dilemmas of the same form, created by the opposing pulls of order and freedom. That a discussion of policy choices affecting standardization of network technologies could be cast in so generic and all encompassing a mould, therefore, should not be thought very remarkable – even though it does constitute a departure from the conventional terms in which economists address the subject. The interesting question to explore is what it is that economic analysis permits us to conclude about how best to mediate between order and freedom in this particular realm of societal arrangements.

From 'order' one may draw predictability, the perfection of performance through repetition and routinization, the economies of simplification and other kindred advantages. Standardization as an intentional act, or as an unintended consequence of the interplay of actions taken for other purposes, creates order by reducing variety. Order offers reduced uncertainty and thus permits economizing on the costly gathering of information; and the establishment of simplified routines is known to be conducive to the attainment and enhancement of many specialized skills. From all these effects will flow immediate gains in productive and allocative efficiency. Such is the conventional wisdom.

There is another side to the matter as many would be quick to point out. Just as people may have inherent preferences for the placidity that accompanies order, so they may have demands for intrinsic novelties as means for combating the malaise of boredom. Some years ago, Tibor Scitovsky (1976) drew an arresting portrait of modern-day, affluent consumers as engaged forever in mixing novelty with repetition in largely vain efforts to maintain a physiologically comfortable equipoise between states of excess stimulation and states of inadequate arousal. Individuals are not the same in their tastes in these respects, nor in the pursuits and material objects that give them satisfaction. We must acknowledge, therefore, that the order achieved by standardization and homogenization can bring efficiency gains only at the cost of suppressing some idiosyncratic sources of consumer satisfaction.

Yet this hardly gives a full accounting of the costs that must be laid at the door of order. The tyranny of systematization and uniformity can be charged also with the stifling of creativity and with constricting the scope for learning and progress via experimentation and the selection of superior variants. It is valid to qualify this charge by noticing that order may serve to focus experimentation in useful directions, and that simplification will often assist learning via the isolation and identification of cause-and-effect relationships. Such effects directly contribute to improving the productive efficiency of established routines. By the same token, the imposition of order must circumscribe what can be discovered: it delimits the sphere of observation and the domain over which selection is able to operate (David and Rothwell, 1993). In reducing diversity, standardization curtails the potentialities for the formation of new combinations and the regeneration of variety from which further selection will be possible. Diversity is the fuel that propels evolutionary adaptation, and, as the biologist Richard Lewontin (1982; p. 151) has nicely put it: 'Selection is like a fire that consumes its own fuel ... unless variation is renewed periodically, evolution would have come to a stop almost at its inception'.

Perceptions of the polar opposition between these two conditions – uniformity and freedom of variation from common norms – has led quite naturally to the conceptualization of a force field existing between them, within which it may be possible to define an optimal situation and alignment for any particular form of public or private policy intervention, organizational structure or institutional design. There has been a long-standing quest for a golden mean, or ideal balance, between freedom and order in political affairs and political economy more specifically. Following on from such habits of thought, modern economists have slipped easily into addressing the question of standardization in regard to technologies and organizational designs by asking how much adherence to uniformity would be 'best', either in the sense of optimal for the economic welfare of the society as a whole, or for various private interests within the whole.

The economists' quest for 'the perfect standard', and for the process, procedure, or institutional form that would reliably deliver such perfection, is analogous to pursuing the grail of an ideal, 'just' condition for humanity – the perfect polity poised between the extremes of individual freedom and social order. Their shared vision of a stationary utopian condition offers an admittedly attractive goal for public policy analysis, but there are many areas of economic life in which the key to beneficial public policy lies in resisting this seductive ideal, or at least approaching it only with considerable caution. The area concerned with standardization policies for information network industries is a case in point.

Although the economic logic of public policy concerning technical standards-setting in general is in an unsettled and confusing state, the present turmoil is especially noticeable in regard to standards affecting the information, computer and telecommunication industries. Nowhere else is one likely to find as strong a consensus in the view that very large public stakes are involved in achieving a greater degree of coordination among the economic agents concerned with these pervasive technologies. At the same time, nowhere else is doubt more widely expressed about the adequacy of relying upon the procedures of the voluntary organizations and institutions (let alone the public regulatory agencies) engaged in developing standards as the principal means of achieving coordination. Indeed, in few areas can one find more uncertainty among the self-professed experts as to the comparative wisdom of leaving coordination to be reached through competitive market processes, in preference to establishing it by essentially political means of consultation and negotiation. Whether we will do better to depend in this sphere upon the workings of invisible hands of market processes, or upon the intervention of visible (public and *quasi*-public) advisers – to employ Greenstein's (1993) felicitous phrasing – is a question that business managers and economists perennially ask themselves, without appearing to be able to arrive at any general answers.

This confused and inconclusive state of affairs shows little sign of being dispelled. That is not because economists are insufficiently industrious in addressing the realities of business and public policy choice, for the literature on compatibility standards, although young, has already grown to extensive proportions. Nor is it the case that the evidence of experience is too meagre to provide grounds for systematic analysis of what works and what fails to achieve coordination in economic networks and industries based on network technologies. Rather, the source of much of the difficulty will remain so long as the discussion continues to be addressed to answering the wrong question. The question of where it is best to intervene to set standards, and where markets are best left to arrive at standards spontaneously (without institutionalized consultation or deliberate public direction and regulatory legal sanctions) if, indeed, they do so, is badly posed for precisely the same reasons that the simplistic application of the metaphor of order and freedom in this area has failed.

It should be clear that I am not faulting the use of metaphor, or 'poetic reasoning' as such. The ambiguity inherent in metaphoric correspondences – such as 'standardization and diversity in technological systems are as incompatible as order and freedom' – may make these essential tools that permit human brains to transcend the confines of logical information processing and generate truly novel thoughts (Johansson, 1988; 1993). However, although metaphors can give us the hopeful, fuzzy maps we need when first venturing into new conceptual territories, where they are followed uncritically for very long, as though they actually were based upon a survey of the new terrain from which routes might reliably be plotted or plans for cultivation might be devised, it is all too likely that one will be led into infertile deserts or treacherous swamps.

The metaphor of order and freedom, unfortunately, turns out to be far more instructive as to the broad nature of the political problems posed in setting technological standards than it is in illuminating the core economic issues, or in indicating the characteristics of the best policy approaches. There are two distinct points packaged in the latter complaint: one concerns incorrectly identifying the nature of the crucial economic problem, while the other concerns the identification of solution strategies. I begin to unbundle them here, but defer trying explicitly to characterize a 'correct' policy approach until the final section of this Chapter.

The conventional juxtaposition of order against freedom in political discourse, historically, has encouraged the simplistic view that the crux of the problem lay in the fact that standardization and technological innovation inherently are antithetical processes. This is the message that emerged, for example, from the classic study by Robert Brady (1933: p. 26) of the rationalization movement in German industry during the 1920s:

> Contrasted with the dynamic, revolutionary, stimulating power of science and technological change, they [standards] represent stability, order, routine and regularity. Standards lose their value with the necessity of change. While they should be established in keeping with the dictates of scientific development, still theirs is the empire of the relatively fixed and changeless in a technically regimentised world.

Brady was led to this striking summary formulation under the influence of what might best be described as the 'philosophy' of the German Standards Committee (Deutscher Normenausschuss) conveyed by the pronouncements of M. Hellmich (1927: p. 25):

> Freedom and order are the two poles between which every controlled development must run. Innate human creative and formative powers crave a condition of freedom; reflective and systematising reason demand order. Both have their justification; only the unfettered interaction between these two coercive forces will give salutary results. Standardization becomes dangerous and harmful when it immoderately narrows the effective range of creative fancy and thereby crushes the most powerful driving force to progress. It dare, consequently, be extended only into realms where development has practically run its course and where the methods and the knowledge, providing a basis for rational ordering and control, lie readily at hand...

The foregoing statements require some qualifying caveats and, indeed, I would argue that it is ultimately misleading for modern policy makers to accept this projection of the 'freedom versus order' metaphor into the arena of contemporary technology standards discussions. This is especially so as regards discussions of the class of 'compatibility' or 'interoperability' standards that are crucially important for the development of network technologies. This formulation is anachronistic in construing the work of standards-development organizations to be inherently directed toward 'uniformitization'. It now has become much more instructive to see the goal of standards-setting in the context of network technologies to be directed towards the attainment of 'coordination'.

SOME SIMPLER ECONOMICS OF COORDINATION STANDARDS AND NETWORK EXTERNALITIES

Seen from the most general theoretical perspective, the subject of 'standards' belongs to the domain of information economics. The establishment of standards *ex ante* has greatest significance when economic agents cannot assimilate without substantial costs all the pertinent information about the commodities that may be exchanged and the processes by which those goods and services can be produced and distributed. Many features that have been noticed as problematic regarding the demand for 'standards', and the supply thereof, are not peculiar to standards. Rather, they are generic attributes of information as a commodity. Lack of super-additivity is one example: just as to have the same bit of information twice does not convey more information than having it once, so having two standards for the same thing does not mean one has more 'standardization', indeed, quite the opposite may be true. This does not mean, however, that there is no extra value in having a multiplicity of standardized channels of communication, since the value of redundancy in information systems is well recognized.

Broadly construed as information, standards may be said to have the economic function of reducing transaction costs. In general, having dependable standards can lower transaction costs by making it simpler for all the parties to a deal to recognize what is being dealt in; and also by limiting practices such as giving short weight, short measure, adulteration, debasement of payment media, erratic rulings in settlement of disputes, and so forth. Private agents' costs of information acquisition obviously can be lowered by the elimination of variety, so 'standardization' (the action of bringing things to a uniform standard) has the effect of facilitating transactions. Elimination of variety also may yield resource savings in production, as is the case when greater uniformity permits economies of repetition, more intensive (larger scale) utilization of fixed production facilities and reductions in the relative importance of set-up versus operating time.

Economists have recognized that standards of all sorts (weights and measures, currency and coinage standards, safety standards, as well as the so-called compatibility standards) can possess the properties that usually are associated with 'pure public goods' (Kindleberger, 1983; Carlton and Klamer, 1983). These properties are: (a) the indivisibility of whatever benefits the good provides among the separate members of the group enjoying them; and (b) the condition that every member of the group has equal access to the total quantity of the good which is available to the group (Samuelson, 1954; Buchanan, 1968).

Public Goods, 'Market Failure' and State Intervention

A contrast may be drawn between standards or social norms that facilitate the coordination of activities among economic agents and goods like novelty and diversity. The former are really public goods by nature; whereas variety and novelty seem closer in nature to pure private goods. Diversities in the form of the variant engineering solutions found for production problems are generated quite naturally during the early development of new technologies. This is when there is still room for experimentation and legitimate disagreement over the best way to do things, as well as over the characteristics of the goods and processes that will be found most desirable at future dates. Furthermore, the egotistical pursuit of fame and fortune can foster deliberate efforts to differentiate proprietary ('sponsored') products and production methods. This includes strategies of creating 'network incompatibilities' and 'switching costs' in order to tie customers who have made complementary investments in the components of a system, the better to be able to extract as excess profits some of the consumer surpluses that they might otherwise enjoy.

The foregoing sources of deliberate diversity can be seen to have been at work in some degree during the early histories of network industries, such as the railways where the variations that resulted in the gauges of track laid initially in Britain and the United States, and the persisting gauge differences on the continent of Europe, are legendary (Puffert, 1991). Some of these same forces reappeared in a new guise during the so-called 'Battle of the Systems' between the Edison-sponsored direct current and the Westinghouse-sponsored alternating current variants of the nascent electricity supply systems at the end of the 19th century (David and Bunn, 1988). They re-emerged once again during the 'Colour TV War' of the 1960s, which saw the French-sponsored Sequential Couleur à Mémoire (SECAM) standard pitted against both the German Phase Alternating Line (PAL) system and the US National Television Standard Committe (NTSC) standard (Crane, 1979; Pelkmans and Beuter, 1986). The static theory of standardization thus suggests that there will be market failure in the form of insufficient standardization which state intervention may need to correct, although the costs at which such correction can be achieved remain a separate issue. The static view of the public goods supply problem applied to the case of information standards is only one part of the story, and perhaps not the most pertinent part.

De facto Standardization as an Uncontrolled Process

The complications have been brought out most vividly by recent theoretical and empirical studies of the economics of standards for compatibility in communication and other network industries (Farrell and Saloner, 1985; Arthur, 1989; Katz and Shapiro, 1986; David, 1992). These have shown that, in the case of a public good like a coordination norm or a social communication standard, competitive market processes also have a propensity to err in the opposite direction: the outcome of private choices in competitive markets may be excessive standardization and coordination of behaviours. Fundamentally, this problem is traceable to the nature of the dynamic interactions among members in 'network' contexts, such as are created by physically linked communications networks; these interactions are characterized by 'positive feedback' effects that cause micro-behaviours to become self-reinforcing.

Let us consider the benefits of technology standardization deriving from economies of coordination in production and exchange. These are found, for example, in the ease with which knowledge gained by purposive Research and Development (R&D) outlays can be transferred among economic units and applied to reduce costs; in the externalities deriving from participation in larger physical networks such as electricity grids; in the savings on the costs of maintaining inventories of spare parts, compared to the levels required where a great variety of makes of cars, or television sets or electric motors, is in use. In these circumstances, the effect of increasing the number of users of a particular variant of a production technology, or mode of exchange such as an organized market, or a contractual form, is to decrease the next choosers' incremental costs of following suit. Therefore, an early accumulation of choices in favour of one variant could create inducements sufficient to lead subsequent agents to select the same variant regardless of their own inherent tastes or predilections in the matter. The result is a dynamic sequence involving historically contingent choices (made either by different actors, or periodically revised by the same actors) which can lead inexorably to high levels of conformity; indeed, to *de facto* standardization on technological or other formulations that by no means need be superior (on technical or economic grounds) to the other, available alternatives (David, 1992; 1993).

Looked at from this angle, when individuals are in social contexts or technological contexts in which they must take account of their interdependence with their neighbours, that is, where their choices are shaped not only by private tastes and information but by externalities arising from the policy commitments that others have made, decentralized decision-making can result in too much standardization, on the wrong standards, arrived at too soon. Thus, we face a reversal of the conventional wisdom based upon the static welfare analysis. There may be a problem of leaving things to the market, not because markets fail to provide enough of the public goods standards, but because market processes can become bandwagons carrying groups of people off in directions that they would not, as individuals, wish to go. This is what is meant by 'lock-in through historical accident'. What brings about 'lock-in' is the positive feedbacks occurring among the community of potential and already committed users. The trouble with such communities is that they are communicative enough to develop bandwagon movements with enough momentum to establish a consensus which will institute even sub-optimal arrangements, but they rarely attain the extreme degree of social communication – the provision of complete information about the intentions of every member of the group – that would enable the undoing of those arrangements once they have become established. Farrell and Saloner (1985) have demonstrated that if a new technological standard (behavioural norm) would be economically superior to the incumbent practices when everyone had switched to it, complete information in the possession of all users would be sufficient to induce everyone to decide independently to make the necessary switchover. This particular route of escape from sub-optimal lock-in depends upon a rigorous backward induction process which leads the last user to switch, given that all others have switched; and the next-to-last to correctly anticipate the decision of the last users, and so on, back to the first user, who will switch in the expectation that all following him will do likewise (David, 1987: p. 227). Incomplete information will readily sever this chain and therefore will prevent it from even starting to form.

The foregoing considerations suggest that the workings of decentralized markets leave much to be desired when it comes to the provision of coordination standards, and more generally, technical standards that facilitate the formation of systems in which there are positive network externalities. There is both a general presumption that private interests will free-ride by not investing sufficiently in the process of developing non-proprietary standards for interoperability, and a prediction that the strategic interests of dominant vendors of network components will incline them to resist choosing compatible designs that would differentially advantage rivals by offering their customers access to a more extensive network. At the same time, the dynamics of bandwagon formation suggest the possibility that market momentum can develop that will result in the premature extinction of a diversity of choice. Prematurity here may take any of three distinctive forms, according to the perspective of the participants involved: (a) selection of a technological system that is inferior to one that is either available, or would otherwise have become available; (b) the reduction of costs of entry to production of standardized products, thereby intensifying price competition to the point that the developers of the successful system are unable to recoup the fixed costs of its creation, hindering their capacity to undertake a following round of technological improvements; and (c) the stranding, or 'orphaning' of a substantial body of users who had adopted network products that failed to become industry standards, and consequently ceased to be supported and further improved.

On each of these counts there may be a good case for corrective public policies, but not necessarily policies that involve governmental intervention to set technical standards (David, 1987). In the case of the first, selective procurement strategies can be deployed to slow the formation of bandwagons, preserving the diversity of technical contenders for market dominance, and thus prolonging the phase of experimentation and radical innovation. Public subsidization of R&D for the development of network technologies and liberalization of regulatory (antitrust) policies in order to foster pre-competitive research collaborations in basic technological advance, would go directly to the heart of the second set of problems. Encouragement of the development of converter and gateway technologies to permit *ex post* compatibility, once the technology has matured to the point that a few rival systems are in contention, would address the plight of orphans and prevent the perceived risks of capital losses through orphaning from slowing the attainment of the critical mass needed to successfully commercialize many network technologies (David and Bunn, 1988; Farrell and Saloner, 1992).

STANDARDS INSTITUTIONS AND THE POLITICAL ECONOMY OF VOLUNTARY STANDARDS-SETTING

Yet another way in which to hinder the formation of market bandwagons which would threaten prematurely to dictate technical specifications to future generations of users might be to encourage anticipations that standards are about to emerge through an alternative channel, a channel that is at least partially insulated from the dynamics of competition for 'installed base', and through which the work of setting standards proceeds at a more deliberate and controllable pace. It is only slightly facetious to suggest that, in the present climate, such might be seen to be the latent economic function of a number of the existing standards-writing institutions – although it most certainly is not their manifest purpose. Nor is it a latent function whose performance is bringing much appreciation to the international standards organizations.

International standards bodies, such as the International Organization for Standardization (ISO), the ITU Telecommunication Standards Bureau and the International Electrotechnical Commission (IEC), comprise a variegated collection of institutions that have long remained inconspicuous and little understood by outsiders. Within the past decade, however, the economic importance of the work of these bodies and their regional counterparts in Europe and elsewhere, and especially their activities in regard to the standardization and 'harmonization' of technologies in the information and communication sector, has begun to be recognized and appreciated more widely among economists (Farrell and Saloner, 1988; O'Connor, 1988; Besen, 1990; OECD, 1991; Swann, 1992). And so they should be, for the formulation and publication of technical standards and the procedures for certifying the conformance of products with such standards, has potentially profound and lasting effects in defining national and global markets, structuring international and interregional competition and patterns of trade and influencing the rate and direction of technological change. The performance of these standards-setting institutions will affect vital infrastructures for the development of the global economy in the coming century. Yet, increasingly they are perceived as being as much a part of the contemporary 'standardization problem' as of its solution.

Institutional Performance Problems and their Sources

The expansion of formal standardization work being carried on by these bodies notwithstanding, members of the business community and some of the technical experts participating in those activities, have expressed an array of discontents with the functioning of the present standardization regime. Excessive delays in the drafting and approval of standards, rising costs, needless uncertainties and confusions caused by the multiplicity of organizations asserting 'jurisdiction' or self-assigning responsibility for standardization in the same or closely related technical areas, alleged biases and lack of due process in the constitution of standards committees, non-responsiveness to the economic interests of under-represented consumers and users of the technologies in question, all have been cited as defects of the process (US Congress, 1992; Dankbaar and van Tulder, 1991; Foray, 1992). The performance of national and international standards organizations alike in this regard has not escaped adverse comment and calls for reform.

Not surprisingly, to these concerns about the performance of existing standards institutions and organizations have been added other considerations registered at the level of national and regional governments. Here, standardization policy is coming increasingly to be viewed as a form of technology policy that may have long-term political as well as economic implications, in addition to its shorter term impact as an element of national commercial and industrial policies. This has been reflected in the formation of regional standards organizations in Europe, such as the European Committee for Standardization (CEN) and the European Committee for Electrotechnical Standardization (CENELEC), as well as the encouragement of various groupings of electronics and computer firms geared toward development of Open Systems Interconnection (OSI) protocols. Policies affecting 'standardization' activities, however, constitute technology policies of an especially complex kind. They have implications that range from the development of suitable modes of collaboration among private corporations in the performance of R&D, to the creation of effective mechanisms for the dissemination of technical information. As more than one observer of the contemporary telecommunication and computer industry scene has pointed out, standards-writing activities have come to involve not just the negotiation of common technical specifications to ensure reliability, connectivity and interoperability, but *quasi*-ideological movements aimed at mobilizing the support of the designers and users of those technologies for the timely provision of such standards (David and Steinmueller, 1990; Barry, 1990).

In the European context this has taken on a manifestly political coloration. The European Commission's initiative toward 'technical harmonization' adopted by the Council of Ministers in 1985, for example, was elaborated in the 1987 Green Paper on telecommunication, and subsequently given concrete institutional form in the foundation of the European Telecommunications Standards Institute (ETSI). The intention is not only to undermine the capability of national standards bodies to deploy technical standards as non-tariff barriers to internal market integration, but also to enlist the participation of technical communities across Europe in forging a 'European technological identity' around a common set of negotiated standards (Narjes, 1987; Pelkmans, 1987).

Many in the standards community contend that, historically, this process has worked well. Nevertheless, there has been a rising chorus of complaints concerning various dimensions of organizational performance. In some cases it is suggested that the open access rules of voluntary standards bodies permit the participation of parties who construe their interests to lie in blocking the issuing of any standards recommendations. However, the issues of representation and 'voice' in these institutions are considerably more complex than this, and require more careful study before importance is assigned to such criticisms. For example, users complain that vendors dominate the standards-writing process, but that may well be the case because they are able and interested in mobilizing research resources and expertise which are persuasive in the context of technical committee deliberations, rather than because the organizational rules and procedures are biased (Weiss and Sirbu, 1990; US Congress, 1992; Foray, 1992). Vendors of equipment complain, in turn, that in organizations where participation is open to individual professionals from the engineering community, the standards-writing process is not 'accountable to industry'. The exclusion of private vendors from some intergovernmental standards bodies, and of small firms and user representatives from effective participation in the work of voluntary standards-writing associations, has raised persisting criticisms of the 'fairness' of the process, and reflects perceptions that standards can be used to cartelize markets and entrench the dominance of incumbents.

The degree of political control exercised over the activities of the ITU by the Public Telecommunication Operators (PTOs) of the major European nations was perceived as a serious obstacle to the movement for liberalization through 'open network provision', which would require the development and implementation of many new interconnectivity standards. Such criticisms are not confined to the long-standing institutional arrangements in the international standards arena; fears are also voiced about the biases that may be inherent in some of the newer developments. To cite one instance: the growing role of regional standards-setting bodies is alleged to favour large transnational corporations, because they can gain access by setting up subsidiaries abroad, whereas most small companies cannot.

Evidently, many deficiencies of the international standards regime have become more prominent as these institutions have been put under increasing strain by transformations that are occurring in the technological and economic environment. It seems less a matter of any breakdown or regime disintegration, and more one of inherent difficulties in (but also some calculated resistances to) adjusting organizational structures and cultures in response to changes in the functional roles that these international bodies are expected to fulfil. Attention has focused upon several profound changes in the global environment which, it is suggested, have significant implications for the manner in which standards are set. For example, the pace of technological change in the ICTs has increased dramatically in recent years. This, in turn, has increased the flow of work for standards institutions. At the same time, it has also changed the nature of standards decisions. Whereas previously standards institutions sought to rationalize technology that was changing only very slowly, or in a regulated and predictable manner, today standards must invariably be set for technologies which are in a state of flux. The changing nature of the standardization process is evidenced by the growing importance of so-called 'anticipatory standards', that is, standards set far ahead and intended to guide the emergence of new technologies, and consequently set far in advance of the markets' ability to signal the features of products and processes that users will value.

The clamour arising for 'standards first', to coordinate the work of many specialists in the design of the components of emerging network technologies has given rise to a new form of standard, which is in reality a 'meta-standard'

(Steinmueller, 1994; Foray, 1993a), that differs from the classic 'product standard' in its intention to describe a flexible architecture encompassing alternative specifications and designs that would fulfil an agreed set of systems functions. The meta-standardization exercise involves a retreat from the premise that expert unanimity is an attainable organizational goal. Inclusive, anticipatory standards-writing is more akin to an exercise in cooperative R&D and indicative collective engineering, engaging the interests and expertise of a widened professional community. It also engages the longer-term strategic interests of a more heterogeneous collectivity of business organizations that must support the research effort. Understandably, strains have appeared from the conflict between the older organizational culture of standards organizations grounded in 'expert engineering objectivity' and the newer corporate culture that views R&D activities as an integral part of a total business strategy.

To tax further the existing institutional regime, some of the confusions created by 'jurisdictional' competition among different standards-writing bodies have their roots in the technological convergence that has been taking place between the once distinct fields of telecommunications and computing. Standards institutions consequently are witnessing great changes in the business constituencies they serve. Industry structures are changing. This, again, is partly a consequence of technological change: the commodification of the computer, for example, eroded the market power and *de facto* standards-setting authority of the once dominant IBM. It is also partly a consequence of regulatory reforms such as the liberalization of telecommunication markets. The intersection and overlapping of technical competences, in turn, has created struggles among different voluntary standards organizations in the 'market' for published standards where there is a temptation to try to meet the rising overhead costs of the management of the committee process by invoking the protections of copyright law and selling standards documents and specifications of conformance testing procedures. When turf battles take this form, the public goods nature of coordination standards is in danger of being jeopardized.

In addition, the growth of transnational corporations, the rise of new centres of technical and industrial competence, and the consequent intensification of global competition has left only memories of that former, more ordered state of economic affairs in which institutions like the ITU International Consultative Committee on Telegraphy and Telephony (CCITT) could be seen to have served effectively in the role of international cartel managers, entrenching the respective monopoly positions of the national PTOs and their favoured equipment manufacturers. For these reasons, it seems, the standards scene in the information and telecommunication industries especially is in a state of flux. There has been growing and widespread dissatisfaction with the long-standing institutional framework for standardization, but no new coherent approach has emerged as ascendant in policy discussions and the movements for reform. In the face of such inertia, new regional standards bodies, such as ETSI, have been brought into existence, embodying new procedures designed to speed the standards writing process in the field of European telecommunications (Besen, 1990). There has also been a turning away from reliance upon formal, *quasi*-governmental modes of standards-setting: numerous private consortia continue to be formed to bypass the formal international standardization process by delivering sponsored standards for *de facto* ratification by the market.

INSTITUTIONAL REFORMS AND THE RELATIONSHIP BETWEEN MARKETS AND COMMITTEES

The re-design of formal standards institutions and governmental regulatory policy mandating conformance with technical standards should reflect an understanding of market failures in *de facto* standards choices. The question of the possible reform or adaptation of the existing international standards institutions and regimes must, inevitably, consider market alternatives to the institutionalized process. The interaction between institutional and *de facto* standards is therefore an important consideration, but the capability of economic analysis for treating such matters remains in a rather primitive state of development. Farrell and Saloner (1988) have contributed a useful beginning by offering an analysis of committees as mechanisms to achieve coordination when participants differ sharply as to which of two mutually incompatible standards they support even though both accept that they will be worse off if they fail to reach any agreement. This analysis compares the coordination of the committee process with that of a simple market leader model, and also examines a hybrid of the two, concluding that the committee takes a larger number of periods (negotiation rounds) to reach consensus than does the market, but tends to coordinate more often (and hence does better, if no value is placed on speed). Farrell and Saloner also find that the hybrid of the two systems performs better than the pure committee process, essentially because the development of alternative paths for coordination in any time period increases the probability of successful agreement.

The results of such theoretical investigations, inevitably, depend upon the structure of the hypothesized game. While the model can illuminate some of the essential mechanisms at work, policy recommendations will turn on the ability to meaningfully compare market and institutional standards-setting in various situations. One limitation of Farrell and Saloner's analysis, therefore, is that they did not specify the respective speeds of their two dynamic

processes in real time. Further, their characterization of committee negotiations as a war of attribution game may be more appropriate to the situation in
which the committee's task is to reduce the diversity among existing technical product specifications, whereas it does not seem apposite to the case of
anticipatory standards-writing, which we have seen to be closely akin to
collaborative R&D.

The development of voluntary standards within an institutional environment cannot, of course, take place in isolation from events in the market. A
firm (or group of firms) may elect, at any stage, to bypass the explicit
negotiation process and seek to go it alone with a *de facto* alternative.
Similarly, once a standard has been negotiated it must also win acceptance
among industrial users or final consumers. Thus, theoretical analyses of
firms' coalitional behaviour in choosing standards is directly germane to
discussions of institutional reform. For example, the cooperative choice of
standards (under the aegis of a formal standardization process) may lead the
firms represented therein to maximize industry profits at the expense of
consumers. On the other hand, when firms choose among standards non-
cooperatively (in *de facto* market standards-setting processes) a subset of
firms may agree upon a standard in order to attract consumers from the other,
incompatible standards-sponsoring firms. Monroe (1993) has shown in at
least one case that individual firms make exactly the choices a social planner
would desire, because their incentives to 'steal' business from one another
are equal in magnitude to the social benefits derived from achieving compatibility. This points to the possibility that the new wisdom regarding the public
policy implications of positive feedback in market-guided processes of *de
facto* standardization stands in need of some qualifications. There may be
market conditions for products with network externalities where government
intervention is uncalled for, whereas participation in formal standards-setting
processes by industry representatives may result in anti-competitive outcomes. Furthermore, *de facto* standards-setting entails a greater role for consumers ('users') in the selection of technical standards than is typically the
case when technical committees and working groups of voluntary standards
organization are given the task.

The most far-reaching implication of the foregoing discussion is that reform of the standards-writing regimes may have to acknowledge that the
need for standards of different kinds in the different phases of a technology's
development, ranging from meta-standards to product standards, gateway
protocols and conformance testing standards, may not be possible to satisfy
within any monolithic institutional structure that has a coherent organizational ethos and internally consistent rules and procedures. The growing
importance of coordination work in the development and implementation of
technologies as we move into the century of elaborated global informational

networks may call for concerted efforts to put in place a more effective institutional infrastructure involving a greater degree of functional specialization on the part of standards-writing entities and, in turn, the provision of mechanisms for closer coordination among them.

CONCLUSION: STANDARDIZATION AND THE DYNAMIC APPROACH TO COORDINATION PROBLEMS

That there are crucial differences between the 'uniformitization for rationalization' and the 'coordination for system-development' interpretations of standardization is evident. The distinction is one that parallels that between defining standards for performance and specifying the precise design and mode of operation as well as the (performance) result. Achievement of complete uniformity may convey certain economic benefits, such as cost reductions through economies of information and scale, and it may be a sufficient condition for the achievement of coordination. Yet, complete uniformity in the specification of all the components of a technological system may not be necessary to coordinate the activities of agents engaged in the design and development of components that will be interoperable, or of improved system configurations that would satisfy users with heterogeneous tastes. Once *coordination* is recognized to be the economically relevant goal, it is quite natural to address a different set of questions. Not content simply to ask how far it is necessary to go in restricting variety in order to secure the benefits of coordination, we also need to know whose activities it is relevant to coordinate by means ancillary to the workings of competitive markets, namely designers, component manufacturers, system assemblers or end users. Correspondingly, it is crucial to determine when in the cycle of industrial development it is most beneficial, and most feasible (which may not be the same thing) to achieve each sort of coordination. These questions are not easy to answer as they involve matters of degree and timing, but they raise the issues that it will be most fruitful for policy makers to confront.

I have argued that the problem of achieving the perfect degree of standardization (read as 'uniformity') is unlikely to admit of a satisfactory solution because it is badly posed. It simply mirrors the formulation of classic welfare analytical exercises that economists have conducted in contexts where simplification of the problem using static assumptions was conventional and did no great violence to the issues of core concern. By contrast, the essence of the questions before us is *dynamic*. The dynamic character of modern technology makes it futile and dysfunctional to view standardization as a purely regulatory matter, to associate the totality of standards policy with the activity of promulgating very specific and stable technical norms to which industrial and commercial practice will be compelled to adhere. This line of thought points to the conclusion that the dynamics of market-mediated standardization processes limit the applicability of static welfare-analytic formulations of public policy problems. We will not penetrate very far beneath the surface of current policy concerns by seeking categorically to decide whether or not government intervention can be justified in promoting the establishment of technological standards. This is so despite the fact that such an approach will draw attention to the existence of divergences between private and public interests in the outcomes of purely voluntary processes, whether they are organized through impersonal markets or committee meetings.

By emphasizing the role of standardization in supporting coordination we are prompted to ask: 'Coordination of which agents for which purposes?' And that leads naturally enough to approaching standardization as an integral aspect of the process of technological development, an essentially dynamic process that cannot be managed successfully without acknowledging its intrinsic uncertainties. Thus, evolving technologies and the market situations of the firms that employ them are highly mutable in ways that preclude arriving at policy recommendations that have a simple, non-contingent and time invariant form. This is the basic orientation that underlies the discussion of the nature of public and private interests in compatibility standards and the mechanisms and institutions that bring them into existence. It does not, however, lead to policy nihilism and counsels of despair. On the contrary, it calls for a more thorough integration of standardization issues in discussions of technology policy. It also points to the substantial scope for improving market performance through greater attention to questions of timing in making selective and flexible public agency interventions, and in promoting reforms and innovation in the institutional infrastructures supporting innovation and diffusion. The need for standardization, and the costs of establishing standards of different kinds, both in terms of the effort of negotiation and selection that is required and of the opportunity costs of the options foreclosed by agreeing on a standard, are almost certain to change over the course of the life cycle of a technology. Variety and uniformity in technological matters must be seen each to have their proper season. The systemic character of ICTs makes it inappropriate to frame policies on the supposition that each constituent element is evolving independently of the others. Rather, the interdependence among the development and adoption processes of technologies that are complementary may occasion the need for standardization in one domain so as to foster the generation of diversity in another.

What is called for is not a static policy stance with respect to the desirability of public intervention, nor an invariant direction in which public action is to be taken. There is no constant 'best form' of standard in terms of the exactitude of specifications, tolerance for performance variations and latitude allowed for adherence. Instead, we might think in terms of finding a 'best policy flux', an optimized path and rate of movement across the mutable landscape bounded by freedom and order; between promoting forms of coordination that support creativity and the generation of variety in the early stages of a technology's development, and promoting coordination in selection and implementation when the technology has matured to the extent that its capabilities of satisfying the variegated needs of users are understood, while attending to the spillover or externalities that such actions may have for interrelated areas of technological development. This latter approach to standardization policies, emphasizing the importance of timing in policy choices, the need for flexibility in altering a policy direction and the instruments utilized to give it effective force, contrasts sharply with the instructions drawn from the metaphor of positioning the polity at the ideal fixed point between the extremes of perfect order and complete freedom. After all, in framing a policy for technology development, one is not framing a political constitution and, accordingly, there is no compelling body of experience that justifies us in instinctively eschewing courses of action and institutional designs that would produce policy oscillations in the space between the poles of authoritarian regimentation and anarchic freedom.

ACKNOWLEDGEMENTS

The support of the Economic and Social Research Council (ESRC) of the UK is gratefully acknowledged. This work was prepared in the context of the Brunel–Oxford Project on the Political Economy of International Standards Institutions, funded by ESRC award L120 25 1003. Dominique Foray and Edward Steinmueller have been responsible for shaping and clarifying my general thinking about most of the issues treated here (to a degree defying adumbration), in the course of many conversations about our individual and collaborative researches on standards and standardization processes. For some material relating to standards institutions (in Section 3, mainly) I have drawn upon unpublished work with Mark Shurmer and Hunter Monroe. None of those who have helped me in these ways should be held responsible for errors and deficiencies found in the result here; nor do they necessarily concur in the views I have expressed regarding public policies towards standardization.

4. Do reforms make a difference? Gearing methodology to assessments of standardization practice

Liora Salter

Standards organizations have been characterized as a 'maze' because no one except the seasoned veterans of the standards community could expect to find their way through the multitude of acronyms and the layers of organizations, committees and working groups. It was often argued that the actual standards, which took many years to emerge and recognized only existing market realities, were ignored altogether, or were even counterproductive. Standards were developed at great cost, but without regard for the market, and without much attention to user needs. They were, in fact, 'technology-driven', reflecting mainly the predilections and technical expertise of the engineers who developed them. Politics, not the need for standards, drove the upper levels of the primary standards organizations.

This is quite an indictment. One is left wondering why anyone ever bothered with *de jure* standardization. It is all the more serious because *de jure* standardization is now commanding more attention than ever before, and all of the standards organizations have undergone major changes in the past few years. Have these reforms made a difference to the actual production of standards so that the critics must now step aside? How would one know now if the reforms have made a difference?

These difficult questions are not likely to be answered solely by economic analysis or by case studies of standards organizations, useful as these approaches have been in establishing an academic perspective on standards (David and Greenstein, 1990; Besen and Saloner, 1989). What is required is an approach that concentrates directly on the environment for standardization: the source and nature of the original expectations, the reasons now for the heightened expectations and the many roles that standards and standards organizations might play independently of their manifest purposes.

This Chapter puts forward the concept of 'policy communities' as the basis of a methodology to bring to light these often neglected aspects of standardization. The approach presented is as developed by Doern and Phidd (1983),

Doern and Wilson (1974) and Doern and Purchase (1991), and considerably extended by Coleman and Skogstad (1990). The data used are drawn from a 4-year-study of Information and Communication Technology (ICT) standardization which, at the time of writing, was in its third year ('The Politics of Communication and Information Technology Standards', funded by a grant from the Social Sciences and Humanities Research Council of Canada – early findings are presented in Salter (1993–94a; 1993–94b)) and from a previously published study of environmental and occupational health and safety standards (Salter, 1988).

The two fields provide valuable points of reference and comparison. In the earlier research on environmental standards it became apparent that the product of standardization was only secondarily the standards themselves, and that there was an explicit connection between voluntary *de jure* standardization and trade. Observations from the current study seem to confirm that this analysis applies also to ICT standardization. In this case, however, it has also become apparent that there is a contradiction at the heart of the standards process. The better attuned standardization becomes to the actual needs of technology and the dictates of the market, the more likely it is that new barriers will be erected to prevent effective standardization. It may well be that the expectations historically attached to standardization were actually furthered by problems in the standards process. Precisely because the standards process has now been improved, new kinds of problems have arisen.

THE STANDARDIZATION ENVIRONMENT AS A POLICY COMMUNITY

The policy community approach defines 'policy' more broadly than government decisions (Coleman and Skogstad, 1990: pp. 14–33). In effect, any resolution of public issues occurring through an open process of decision making is considered to be policy. The 'community' consists of all the organizations, government departments, *quasi*-independent bodies, industry associations, firms, advocate groups and individuals whose actions have a bearing on the resolution of issues. The policy community approach examines the environment within which issues are defined initially and later resolved, and the roles played by various groups in reaching this resolution.

Policy communities are not bound to national communities – public issues can be resolved in any national or international political context. Invariably, national governments play some role, but industry, labour, media and public advocate groups are also 'members' (or potential members) of a policy community if their actions have a bearing upon the resolution of the issue under consideration. In some cases, these members are readily apparent, while in others their presence can only be discerned by careful study. The resolution of issues can also take many forms. It can involve a formal decision (which may or may not be effectively implemented and enforced) or simply a resetting of the agenda for some or all of the members of the policy community.

ICT standardization provides an excellent example of a policy community. Indeed, it should be understood as a collection of sometimes overlapping and interconnected policy communities, each reflecting different technologies such as telecommunications, radio and computing. Governments always play a role, sometimes even the pre-eminent role, but the resolution of the issues in standardization need not result in a government decision, that is a mandatory standard (Salter, 1993–94a;1993–94b). A public decision-making process is involved, although it need not attract much public participation nor conform to democratic conventions.

A decision in the form of a standard might well result from the activities of 'members' of the standards community, but there are numerous cases where standardization activities have been carried out over long periods of time without producing anything. Often the activities seem to have been directed towards avoiding the creation of a standard. Membership in the policy communities represented by standardization is often difficult to ascertain directly, although most standards organizations have formal rules for admitting participants as members. 'Members' in this case include all the groups, associations, firms, governments and individuals who play a role, directly or indirectly, in shaping both the process of standardization and its results.

Already it should be evident that when applied to the study of standardization, the policy community approach does not centre upon standards *per se*. It cannot provide specific answers to questions of whether standards are beneficial, whether they foster or impede innovation, whether standards organizations are cartels in disguise, or whether standardization operates in the public interest. These are questions more properly addressed by economic analysis. Nor is the policy community approach likely to deliver detailed descriptions of the organizational mandates and structures of standards organizations, or of the standards process in a single country. Its primary focus is not on institutional arrangements as such, but on the *whole environment* within which standards emerge – an environment that is studied in several different ways.

Policy ideas

Firstly, it is important to locate the prevailing ideas that generate expectations and justifications for the actions taken by 'members' of the policy community (Coleman and Skogstad, 1990: p. 17). In meetings of standards committees in the International Telecommunication Union (ITU), for example, continuing reference is made to 'deregulation' and 'liberalization'. Both are said to be driving forces in telecommunication standardization, notwithstanding the fact that in many ITU member countries neither deregulation nor liberalization has progressed significantly. Furthermore, even in the advanced industrial countries, each national case is sufficiently different so that a single term, like 'deregulation', is not enlightening. In studying standardization from a policy community perspective, both 'deregulation' and 'liberalization' are treated as ideas, rather than as necessarily reflecting actual changes in the monopoly status of many telecommunication industries.

Although the ITU is a treaty-based organization bringing together national delegations headed by Government administrations, only in a few delegations are these administrations actually directed by their industry participants. So powerful are the ideas of deregulation and liberalization in the ITU context, however, that they have become accepted as accurate descriptions. Moreover, because they are accepted as accurate, actions are consequently taken on the basis of these ideas.

Lest this point be misinterpreted, it should be emphasized that significant pressures do exist for both deregulation and liberalization, and that many key countries in the ITU have embarked upon policies in support of both. The problem is not that members of a policy community embrace an ideology with little applicability to the actual circumstances in which they operate, but, on the contrary, that ideas, such as deregulation, have a life of their own. It has long been understood by social scientists that an idea accepted as factual will have real consequences and, furthermore, that some are 'keywords' – seminal in their capacity to orient a community towards shared values (Williams, 1976).

Policy Institutions

The term 'institution' does not imply that policy community analysts will conduct studies of specific organizations, when the actual scope of their interest is considerably broader. From a policy community perspective, 'institution' refers to the constellation and interaction of organizations, each with its own mandate, rules and conventions. It also refers to the market, which forms the environment within which standardization operates, and to the network of trade and industry associations through which various members give effect to their ideas, needs and predispositions. (Doern uses the term 'structure' rather than 'institution'. Both terms are problematic in the literature, but 'institution' now has a useful reference in international relations as well as policy analysis; see, for example, Ruggie (1993)).

Emphasis on institutions is especially useful in the study of ICT standardization. Participation by members cannot be accounted for simply in terms of their interests. It is impossible to find direct correspondence, even in terms of attendance, between members' interests and their actions. As critics of standardization so rightly have pointed out, the standards community is a maze. At one moment, members will participate in one organization as representatives of their employer/industry interests, and at another moment as officials representing national policy in another organization. They will call themselves both 'standards developers' and 'users'. Depending upon the issue, they will attend meetings as individuals, as representatives of a trade group, or as representatives of yet another standards body. Finding a single identifiable interest for each 'member' from among this complex of potentially conflicting interests is all but impossible. It is much more useful to concentrate on the complex of institutions that generates the incentives and constraints, bringing different interests to the fore in different circumstances.

Policy Processes

From a policy community perspective, a third level of analysis is critical – one that focuses on the processes through which decisions are made. This necessarily involves looking at the legislative and regulatory frameworks, the formal mandates of the participating organizations and the formal decision-making rules. At the same time, the policy community approach also places emphasis on the informal rules and conventions which can, on occasion, be of equal or more importance. Furthermore, the analyst examines the relationship between the formal or legislative rules and the informal or practical methods of arriving at decisions.

All standards organizations are strictly governed by formal organizational mandates and decision-making rules. Rules govern who speaks at meetings, the appropriate scope of activity for each committee, and the approval of recommendations. A great deal of attention is paid to the distinctions between mandatory and voluntary standards, and between treaty-based organizations where governments are the primary spokespersons, and consensus-based organizations where government officials function as one among many types of members.

Yet legislation, regulations, rules and bylaws are considerably less important than they seem. In some delegations headed by government, for example, industry members write all the contributions and may even script the oral comments and replies from the delegation. Furthermore, treaty-based organizations regularly set voluntary standards and consensus-based organizations can produce standards that are adopted later as mandatory. Understanding the actual (as opposed to the formal) decision-making rules requires an appreciation of the fact that within standards organizations there may be many distinct levels of operation, each with its own conventions.

THE MEMBERSHIP OF A POLICY COMMUNITY

Not all members of a policy community participate openly, even though a public or open process of decision-making may be involved. Some exert their influence behind the scenes, and it is the task of the policy community analyst to locate such members and identify their roles within the policy community (Coleman and Skogstad, 1990).

Members of the standards community often draw attention to the openness of the standards organizations. This seems to suggest that distinctions between members and non-members of particular organizations are of little consequence and that efforts are being expended to recruit new members from a wider community. In practice, however, the distinctions drawn between members and non-members in this community are critical. In most standards organizations, recruitment efforts are minimal and, just as in a geographic neighbourhood or community, there are implicit criteria to determine who belongs and who may claim the right to exert influence.

Notwithstanding the frequent references to openness and inclusion of various new groups, many barriers restrict membership. The standards community is awash with acronyms, which change constantly and cannot be assimilated without long association. The debates make no sense unless one is familiar with the language and acronyms in which they are conducted. The procedural rules are complex and frequently changed, so that even long-standing members have difficulty determining the appropriate way or place to make contributions. Committees produce a voluminous paper record of every contribution, draft proposal, proposed decision and decision (a barrier to participation in itself). Some of the members are comfortable with the documentation and discourse, while other members lack the background necessary even to understand the debates.

It would seem, then, that under these circumstances, informal networks take precedence over formal criteria in determining the actual nature of membership. Even this picture is misleading, however. Informal networking has its limits in standards organizations, because participants must cover their own costs. Control is exerted over who will participate in the standards community in that participants are normally sponsored by firms or governments.

Associational Linkages

Coleman and Skogstad (1990) argue that several distinctions should be made among members of any policy community. The first refers to whether members are oriented towards policy advocacy or policy participation. Advocates function primarily outside the process, exerting pressure on decisions through direct and indirect lobbying. By contrast, participants are directly integrated into the fora where decisions are made. This distinction is important to members of the standards community who routinely distinguish between those who write contributions and those who exert influence through their status as users of standards or technologies.

Coleman and Skogstad (1990) also make distinctions in terms of the level of integration among the various members and their representative organizations. Thus, for example, an industrial sector might be considered to be tightly integrated because its members are represented through strong central organizations. Alternatively, a policy community might be characterized by loose associations among its various members, each member concerned primarily with furthering its own interest. The nature of the associational bond is of paramount importance in studying standardization. This is true not only because each standards organization is itself an association which can be tightly or loosely integrated, but also because those who participate might well reflect the general interests of the industry as a whole in the context of one debate, and the specific interests of a particular firm in another. In other words, the interplay of general and specific interests in each organization, and with respect to each standard cannot be determined *a priori*.

In fact, affinities and associations among the members in the standards community can be quite difficult to determine. In the telecommunication sector, for example, some government administrations see their role as one of coordinating and facilitating contributions from industry, whereas others see themselves as the proper, and perhaps even the only, legitimate representative of the public interest. In other words, the 'government administration' is not one category. Sometimes standards organizations make distinctions between the industrialized and developing countries. At other times, more significant distinctions are drawn between the highly industrialized countries capable of making many expert contributions to standardization, and those with relatively few independent expert contributions to offer. Furthermore, many of the firms involved are multinationals, active in more than one standards organization. How their nationally based experts render opinions, form alliances and coalitions and define conflicts independently of each other without undermining each others' interest or the interests of the multinational as a whole is in itself worthy of study.

In a policy community approach individual members are seldom dealt with in their own terms as bearers of interests and thus as participants in interest group negotiations. More attention is paid to the multiple relationships among the members, many of which do not emerge during interest group negotiations. The policy community as a whole is regarded as a highly dynamic situation within which the formation of alliances, coalitions and the emergence of conflict are continually taking place.

POLICY COMMUNITIES AND POLICY NETWORKS

The term 'policy community' refers to the total environment within which policy decisions are made. However, each controversy or specific decision involves different members, institutions and relationships. A 'policy network' is the pattern of communication among particular groups of members emerging around each controversy or decision. Policy networks are subsets of policy communities, brought into being around specific decisions.

Thus, if one were to speak of the policy community in forest management, one would explore the relationships among all of the industries, government departments, regulators, advisory committees and advocate groups. With respect to a decision to permit logging in a particular territory, however, not all of these groups would be active and, moreover, the specific relationships among them take shape within the controversy at hand. Policy networks are studied on a case-by-case basis.

Anterior Factors

There are two ways of studying the emergence of policy networks. One focuses on anterior factors. This is familiar public policy analysis and requires examination of broad economic and political developments in terms of their effects on particular controversies or decisions. In the case of standardization, this kind of study is relatively straightforward.

The situation with respect to *de jure* standardization has changed dramatically in the last decade and the changes are in evidence in almost every standards controversy. Privatization has lifted standardization out of the domain of single government departments, whose mandatory standards were developed in the normal course of their operations. Liberalization brought new *de jure* standards organizations into being. In any newly competitive environment, some entity must be responsible for ensuring that justice is seen to be done in trade disputes, many of which revolve around issues of technical standards. Deregulation has made it necessary for some entity other than government to play the role of coordinator and referee, and, consequently, it has enhanced the stature of *de jure* standards organizations, and has also been a catalyst for reform in these organizations.

The formation of regional trading alliances is dependent upon the establishment of mechanisms to ensure trade relations and harmonization within the region. Differences in technical standards cannot be allowed to constitute barriers to trade. (For further discussion of the relationship between standards and trade issues consult Nusbaumer (1984), Middleton (1980) and Strawbridge (1991)). The same situation may be true for the General Agreement on Tariffs and Trade (GATT). GATT provides an excellent example of how external developments (in this case, the beginnings of a world trade agreement) can place standardization at the centre of trade relationships. It is also an example of the heightening expectations of *de jure* standardization, which proceeds quite independently of the successful conclusion of standardization negotiations or of the capacity of existing standards or standards organizations to meet those expectations.

Viewing the contemporary economy as dependent upon innovation and integration, standardization becomes an integral part of a successful response to globalization. From this perspective, coordination among firms is crucial and mechanisms to effect coordination are highlighted. In this context, *de jure* standardization, no less than strategic alliances, represents a way to combine the dynamics of market-based competition with the requirements for coordination.

The ICT sector, for example, is highly dynamic, with significant development costs and short product life cycles. Furthermore, these technologies underpin innovation and change in other sectors. For these reasons, attention must be paid to standardization, especially if, as is increasingly the case, *de jure* standardization can be done at the precompetitive stages of R&D and perhaps even in conjunction with newly formed strategic alliances and research consortia.

In other words, *de jure* standards have been cast into high profile for reasons only indirectly related to their benefits or disadvantages for individual firms. Attention is now focused in an unprecedented manner on the role played by *de jure* standards in a world increasingly characterized by privatization, deregulation, liberalization and regionalization. The pervasive argument is that wealth generation depends upon technology and innovation and that industrial strategies should be based on standardization instead of subsidies, tariffs and regulation.

Studying the Internal Dynamics of Policy Networks

Most of the key relationships in policy networks are not explicit, nor are there easy ways to gauge the importance of any change. For example, how do changes among the sponsoring firms, such as the formation of strategic alliances, alter the prospects for standardization? In theory, strategic alliances and standardization ought to be mutually supportive, because both involve erstwhile competing firms in relationships directly related to the precompetitive stages of technology development, and both depend upon a high degree of cooperation. In a recent exploratory study of alliances and consortia in Canada, however, little support was found for *de jure* standardization on the part of consortia and alliances (Ruby and Salter, 1993).

The policy community approach does not provide much guidance in answering questions like this. One can rely upon archival materials and 'elite' interviews, but neither method provides an inside look at the dynamics of policy networks as they evolve over time. Other methods are required, and in this case, they are best drawn from outside the policy community literature. Three are introduced here, ethnography, discourse analysis and a technique which I have called (for lack of a better term) 'signal events'.

Ethnography

Ethnographers treat the standards community in the same manner as they would any other kind of community. They use similar methods to those employed by anthropologists in studying hunter/gatherer cultures or by sociologists in studying urban subcultures. Researchers seek membership in the community of interest, observe its kinship and other patterns of association, identify its values, document its language, and examine its artefacts and practices. What is required is a long-term commitment to participation and techniques for the systematic observation of seemingly unruly and unpredictable situations (Burgess, 1984; Agar, 1981; Cicourel, 1973; Burgess, 1981; and Spradley, 1979).

In studying policy networks, ethnographic techniques are especially useful in identifying patterns of association and communication. The purpose of sustained systematic observation of the standards community, done from within the community as a participant in the various networks, is to understand what gives this community its integrity, and what gives each of its networks coherence and influence.

Discourse analysis

The policy networks in standardization are especially apt locations for applying discourse analysis (Hawkins, 1992a). Much of the activity in standards meetings consists of 'wordsmithing' – the resolution of conflicts by the selection of the particular words to be used in a standard. Any so-called 'incorrect' choice of words leaves the various members in conflict with each other. By contrast, a 'correct' choice connotes agreement and thus a standard. In standards meetings, many hours will be spent choosing correct words, even when the result is an awkward construction according to the normal rules of language.

In this case, the correct word is one which allows members with different interests to see their views reflected in the same standard without necessarily compromising their interests. For example, after much debate, the phrase 'are expected to' might be replaced by 'should' because members believe that the use of 'are expected to' implies a mandatory standard, whereas 'should' implies a voluntary standard. For those who are preoccupied with avoiding mandatory standards, this debate serves as a means to underscore the purpose of their activity. It permits their agreement, where it otherwise might be precluded by caution.

That an outsider might draw none of the inferences in this example or see any threat of mandatory standardization whatsoever from the use of 'are expected to', is beside the point. For members of the standards community, inferences are always drawn from the wording of standards, and consequently each word is debated as if the choice of words were crucial. Moreover, within the different policy networks in the standards community, the same words take on different connotations.

Signal events

'Signal events' are those conflicts, moments or decisions in the debate which occur unexpectedly and which illustrate the otherwise hidden dynamics and relationships within a policy network (this approach is similar to that in Cicourel (1973)). Signal events occur where the normal and unexceptional routines break down. By virtue of having to be re-established or re-negotiated, the original routines or assumptions are exposed. Other types of signal events occur either when conflicts surface or when other actions are taken which cannot be explained by reference to straightforward observations, or even to the interests or political predilections of those involved. For the analyst, these events signal the existence and importance of something other than the immediate situation. The task of the analyst is to render these events intelligible. Three examples illustrate the significance of signal events in standardization.

The first is the current controversy regarding Intellectual Property Rights (IPR). It is an excellent example of a signal event occurring when the normal routines have broken down. IPR has always been at the heart of voluntary standardization, with IPR owners agreeing to cede their rights under certain conditions. Why, then, has IPR become publicly controversial in standardization today? And why is it the subject of intense debate in some standards organizations while remaining off the agenda in others? It would seem that there has been a change in the nature of standards, or perhaps in their significance, that has increased the importance of what is being decided for those with IPR at stake. Signal events are of great assistance in identifying the important questions for further research.

The second example concerns boundary disputes (Jasanoff, 1992). The organizational charts of the various standards organizations, and the questions assigned to the technical committees suggest that the division of work within the standards community is almost self-evident. An overview of the standards process could also easily be prepared showing each organization to have a distinct role in support of others. Thus, it is a signal event when the allocation of work items, and the relationships among technical committees and between the standards organizations is highly problematic and controversial, when boundary disputes are increasing, not decreasing, in spite of organizational reform.

In the ITU, for example, a reorganization has recently taken place affecting both telecommunication and radiocommunication. All standards questions pertaining to telecommunication and radio have been assigned to one department in the ITU, while matters concerning issues related to developing countries or radiocommunication have been placed in different departments. It would seem relatively straightforward, then, to reassign work among the departments in terms of whether or not standards are involved, particularly in the light of the fact that considerable duplication of work had existed previously and that telecommunication and radio technologies are rapidly converging. In fact, the reallocation of work has been slow, difficult and highly controversial. Proposals to reallocate work necessarily alter the balance of interests and necessitate the renegotiation of boundaries. They call into question the membership criteria within both the established and the new spheres of influence and interest. Obviously this is a matter of considerable significance to those involved. Studying how these controversies are resolved is crucial to understanding the policy networks in standardization.

The third example of a 'signal event' concerns the reference to 'users' (Tamarin, 1988; Anderson, 1991; Allen and Gilbert, 1993; and Dankbaar and van Tulder, 1991). At first glance, emphasis on user participation is to be expected if standards are to become increasingly market driven. It might also be expected that scepticism would be expressed about the feasibility of involving users, whose interests are seen to be too diverse and diffuse to be reflected properly. As a result, it is also not particularly surprising that there is disagreement about the value and feasibility of user participation, nor does this situation constitute a signal event. What turns it into a signal event is the difficulty encountered by those engaged in standardization in coming to any agreement about how to define the user. If a position can be taken on the desirability or feasibility of user involvement, it should be clear whose presence is at issue. Yet the definition of 'user' is frequently debated and changed.

THE POLICY COMMUNITY APPROACH AND ASSESSMENT OF THE IMPACT OF REFORM

How should one gauge the effect of reforms? The standards community is highly dynamic and complex. An overall picture of the importance of standards might easily be belied by the actions of a single group within that community. The relationships among standards organizations, and among their various products can be as important, or more important, than the actual standards. Each standards organization is best understood as having a number of quite separate levels of operation, each subject to different formal and informal decision-making rules.

Within this complex, membership distinctions, organizations, work, mandates and so on derive importance from the great number of boundary disputes which occur. These disputes suggest that the distinctions need to be renegotiated continually. This comprises a significant part of the activity within the standards community. To deal with this observation, it was suggested above that the standards community is made up of many spheres of interest, each reflecting the preoccupations and prerogatives of its members.

These various spheres of interest and influence appeared relatively stable for many years. Recent pressures – new trade alliances, changing economic paradigms, forces of deregulation and liberalization and so on – have combined to render their boundaries highly controversial. In effect, reforms within the standards organizations are both outward manifestations of pressures forcing renegotiation of the spheres of influence and interest in standardization, and pressures in their own right.

The standards community is characterized by policy participation, not policy advocacy. To influence this community it is necessary, in effect, to join it on its own terms. Significant barriers are erected against the type of intermittent or partial commitment that would be appropriate from small or medium-sized manufacturers, newly industrializing countries or from users of the technology and standards.

Within the standards community, the power of ideas – deregulation, privatization, the preeminence of international organizations to name but a few – and of language should not be underestimated. Notwithstanding the degree to which any idea or term reflects an accurate description of current realities, the debates about ideas and language serve as media for the dissemination of influence. Whatever its importance with respect to producing standards, the standards community offers an important venue for exerting influence upon public policies. Equally, it offers its dominant members a venue for influencing other industry members. If the original criticisms of *de jure* standardization are taken at face value, it would appear that the main benefit to be gained from participation in the standards community has been, and may still be, closely linked to the patterns of influence established through this community.

There is some evidence that the situation has now changed, and that both standards and standards organizations are becoming more significant in their own right. What is not yet clear is whether the reforms undertaken by virtually all standards organizations are the catalyst for change or, more likely, simply its by-product. Whatever its source, however, the changes have generated considerable churn within and among standards organizations.

Whether or not the status of the actual standards has changed, and regardless of the effect of the reforms, the normal routines which sustained the standards community for many years will no longer suffice. The long-held views which sustained participation have now been questioned as the analysts of 'signal events' so clearly suggests. New justifications for participation have also been made and they have been offered by members of the standards community to respond to the heightened new expectations.

It is far too soon to tell whether the reforms will be sufficient to allow standards organizations to respond effectively to these expectations, or whether policy advocates will take their requirements outside the standards community and perhaps away from *de jure* standardization altogether. Organizational reform is not the central issue – it is one among many necessary steps if the new expectations are to be met.

That standards organizations will continue to exist is probably not at issue. Standards organizations had a long history prior to reforms, indicating that, whatever their capability in developing standards or in responding to industry needs, they have served an important purpose for their participants. What that purpose now should be, and who should function as members of the standards policy community, is precisely what is at stake, not just in the reforms, but also in the complex interactions now taking place within and outside this community.

5. Stakeholder involvement in the administration of environmental standards

William Leiss

INTRODUCTION

Debates about how to set appropriate and enforceable environmental standards and, more importantly, how to enforce compliance with them thereafter, have been going on since the 1970s (Victor, 1979). Since the beginning of the 1980s, however, economic and regulatory aspects seem to have dominated the debate, with a dichotomy between 'command-and-control' versus 'incentive-based' approaches to enforcing regulatory compliance being the leading theme. These aspects also at first appeared to pit business interests against environmental or public interest advocates, with the former favouring market incentives and the latter the issuance of government edicts (with their overtones of punishment for socially irresponsible behaviour).

As Stavins and Whitehead (1992) have argued, the 1990s have brought widespread recognition, almost everywhere in the industrialized world, of limitations on the scope of public sector activity and, simultaneously, of greater pressures on national economies from competitive forces on a global scale. These factors have led to a greater awareness of the costs of regulating and mitigating environmental impacts and a keen interest in the comparative cost effectiveness of different ways of meeting environmental standards. This in turn has shifted the perspectives of at least some participants in those earlier debates, yielding enhanced support for the position that what matters most is achieving a high level of environmental quality by the most efficient means available while, at the same time, doing least harm to the competitive position of the national economy (Carnegie Commission, 1993).

Yet there are other dimensions, apart from economic ones, which are relevant to the simultaneous attainment of an acceptable level of environmental quality and material well-being. These relate to citizen participation in decision-making and are easier to conceptualize when they are put in terms of the concept of 'risk' (Krimsky and Golding, 1992). Setting and enforcing environmental standards are in a sense just one form of health and environmental risk management (Morgan, 1993). In risk management, the substantial economic benefits which accrue from industrial activity are balanced against the threats to health and ecosystem viability to which that activity gives rise, through a decision-making process which is perceived to be comprehensive (taking all relevant factors into account), equitable (in terms of distributive justice) and respectful of commonly accepted values (in terms of human dignity and the duty of care we owe to non-human entities). In operational terms the two key concepts are: (a) risk/benefit trade-offs, where risk managers try to give a quantitative estimate for each side, and then to see how each measures up against the other; and (b) acceptable risk, which may be seen as a threshold of harm above which no trade-offs are permissible, no matter how rich the fund of benefits.

The easiest example of an acceptable risk/benefit trade-off will be found in the area of prescription drugs, where it is the same individual who assumes the risk (adverse side effects) and reaps the health benefit. To take an environmental example, there is widespread public acceptance of the estimated adverse health and economic consequences from air pollution levels attributable to automobile exhaust, as compared to the benefits of automobile use, and the trade-off (whether explicit or implicit) may be regarded as being broadly acceptable, despite the fact that, for example, there may be clear inequities in the distribution of risks and benefits. So far as acceptable risk itself is concerned, in many countries the risks associated with nuclear power generation are broadly regarded as being unacceptable, despite the overall experience to date with nuclear reactors and the wide margin by which benefits exceed risks in the opinion of most 'experts'.

Achieving a level of environmental protection which yields a sense of acceptable risk and acceptable trade-offs is in practice an exceedingly difficult affair, which explains in part why risk management decisions are often intensely controversial (Leiss and Chociolko, 1993). The difficulties arise largely for two reasons: firstly because of the inherent complexity of ecosystem processes (so that managing environmental impacts involves a huge number of relevant factors); and secondly, because of the inherent uncertainties in the characterization of those factors (which are usually expressed as probabilities of outcomes). Thus, despite the immense talent and resources devoted to the scientific description of environmental impacts, on which standards are expected to be based, the existing scientific basis of much environmental regulation is suspect (M'Gonigle *et al.*, 1995); in addition, pressures arising in the bureaucratic and policy structures of government subtly bend the scientific knowledge base towards policy-directed outcomes (Salter, 1988). Compounding these difficulties is the lack of trust in both governments and industry, expressed by much of the public in opinion polls, with respect to the activities of both in protecting the environment.

These very circumstances make broad citizen or stakeholder participation in risk management decision-making more, not less, imperative (Renn, 1992). In this Chapter, it is argued that, at the most basic level of administering the process for enforcing day-to-day compliance with environmental standards, routine involvement of key stakeholders is indispensable for ensuring that community values governing acceptable risk and acceptable risk/benefit trade-offs are upheld. Finally, it is shown that the process of stakeholder involvement fulfils the demands for cost-effective and economically viable environmental protection. To set the stage for these arguments, the debate about the two types of approaches to enforcing compliance with environmental standards is first reviewed.

ENSURING COMPLIANCE WITH ENVIRONMENTAL STANDARDS

The literature distinguishes between what are called command-and-control approaches and incentive-based approaches to ensuring compliance. Conventional command-and-control regulatory mechanisms are of two types (Hahn and Stavins, 1991):

1. uniform technology-based standards, which 'identify particular equipment that must be used to comply with a regulation. For example, utilities may be required to install flue gas scrubbers to control sulfur dioxide emissions or electrostatic precipitators to remove particulates';
2. performance standards, which 'typically identify a specific goal (for example, maximum allowable units of pollutant emitted per time period)' without specifying the means for achieving that goal.

On the other hand, incentive-based approaches include five major types: pollution charges, marketable permits, deposit-refund systems, market barrier reductions, and government subsidy elimination (Hahn and Stavins, 1991):

1. pollution charges: taxes on emissions, which, among other things, offer incentives for firms to develop and install progressively better pollution control technologies;
2. marketable (tradeable) permit systems, where an allowable overall level of pollution is set by government and is then divided among producers, some of whom may sell or lease their surplus permits to others;

3. deposit-refund systems: well-known for such consumer items as bottles and cans, but also feasible for hazardous waste (used motor oil, batteries, vehicle tyres and so on);
4. removal of market barriers: for example, voluntary exchange of government-granted water rights among owners to facilitate efficient use as is now done in Southern California; and
5. elimination of government subsidies: for example, excessively low stumpage fees on crown-owned land in Canada, or public land grazing policies (US), or agricultural price supports which sustain economic inefficiencies or underprice natural resources.

The debate which deals with the claim that incentive-based schemes can achieve adequate environmental protection better than command-and-control approaches from the standpoint of cost-effectiveness is not addressed here. Even the defenders of this claim concede that a mix of the two approaches is required (Hahn and Stavins, 1991; Hahn, 1993). However, many jurisdictions are now rewriting legislation to provide for enhanced responsibilities for corporations and their individual officers – as well as an expanded definition of owner to include trustees, lenders and receivers – with respect to remediation of polluted sites, all of which have 'given companies substantial incentives to comply with environmental guidelines, orders and standards' (Kentridge, 1993).

Interestingly, these sources tell us that the monitoring and enforcement aspects of environmental standards are indispensable parts of a compliance strategy and that these must be the responsibility of governments. It is argued in this Chapter that it is precisely these aspects which are best facilitated by stakeholder involvement.

INVOLVING STAKEHOLDERS IN ENVIRONMENTAL RISK MANAGEMENT

In a fascinating passage, two of the leading authorities in the largely economics-centred debate about cost-effective approaches to environmental protection turn the discussion towards another dimension entirely: 'An added benefit of incentive-based approaches is their tendency to make the environmental debate more understandable to the general public. These approaches focus attention directly on what our environmental goals should be, rather than on difficult technical questions concerning technological alternatives for reaching those goals' (Hahn and Stavins, 1991: pp. 13–14). The full implications and importance of this insightful comment have not been sufficiently explored to date.

The main point of this remark is that citizens have a direct interest in the concrete outcomes of actions taken to enforce environmental standards, and only an indirect or derivative interest in the purely technological solutions through which those standards are satisfied. Yet this is just the beginning of wisdom on this subject, and it is precisely the reconceptualization in terms of risk that enables us to go further along this path, keeping in mind that the bottom-line determinants are acceptable risk and acceptable risk/benefit trade-offs.

Upon examination, there are a great number of key micro-management determinations – or, in other words, exercises of discretionary judgement – in every risk management decision which, in the absence of stakeholder involvement, will be made behind closed doors by the bureaucracies working for industry and governments. Some of these decisions are about whether or not to allocate scarce resources to a risk assessment at all; some relate to the degree of credibility of the scientific data which have been accumulated; and some involve the making of trade-offs, in an arbitrary fashion, which may or may not prove to be publicly acceptable. A partial list of such micro-management issues follows, arranged according to the received schemata of risk management decision-making.

Risk Characterization

- Hazard identification: are there any toxicological studies at all and, if so, are they credible?
- Exposure assessment: are there any reliable human exposure data at all? Are the pharmacokinetics of exposure known?
- Have populations at special risk been taken into account?
- Extrapolations from animal studies to estimated human health effects: have these been done reliably?

Risk Assessment

- Have the correct weights been assigned to the various factors in hazard and exposure that will be combined here to yield the risk value?

Benefits Assessment

- How reliable are the claimed economic benefits?
- Have all the hidden costs been identified in arriving at a net benefits figure?

Risk/Benefit Trade-offs

- What is the basis for comparing adverse health effects in terms of illness and death, on the one hand, with monetary benefits, on the other? Has the implicit economic value of life and suffering been identified?
- What are the distributions of risks and benefits among various groups in the population?
- Is any compensation to be offered to those who are exposed to excess risk?

Risk Management

- Who is responsible for monitoring and enforcement, and what instruments and resources are available for these purposes?
- What cost-effective options are available for risk reduction?
- Has the risk been communicated adequately to those affected?

In Canada, when the decision is made, it will take the form of a national environmental standard. For example, in ambient air in occupational settings, allowable concentrations of formaldehyde, a very useful industrial chemical used in many consumer products, will be one part per million.

The above list is a shorthand for the kinds of questions which may be asked of a risk management decision. Indeed, most of them are highly technical in nature. There is evidence to suggest that ordinary citizens can make reasonable judgements about such matters (Kraus *et al.*, 1992). The questions which interest stakeholders most are qualitative in nature and may be divided into three types:

1. How complete is the data set, and how important is what is missing?
2. Can the numbers which do exist be 'trusted'?
3. What are the value judgements being used in the calculation of trade-offs?

Although these questions refer to matters which are expressed in technical terms, the questions themselves are not about technical matters at all, but rather about the 'human meaning' of the technical data in which risk assessments are couched. A broad range of stakeholders has an interest in these three questions. Moreover, the simple fact of their being present in the room when these questions are posed under circumstances where the institutional risk promoters (governments and industry) are compelled to try to answer them publicly, is itself a contribution to the dignity of the participants as citizens in a civilized democracy. However, as indicated in the following section, it is often possible to go beyond this, and to negotiate an outcome where at least some progress has been made towards improving the outcomes of the trade-offs for those who will bear the greatest risks.

Other factors increase the relevance of stakeholder involvement in risk management decision-making. Firstly, changes in legislation which impose liability for remediation on lenders, receivers and trustees in bankruptcy (Kentridge, 1993) in effect increase the number of key stakeholders and will tend over time to establish higher levels of public accountability for environmental impacts.

Secondly, there is the relevance of comparisons between two competing models of enforcement behaviour for regulation – enforced compliance, that is, limited bureaucratic discretion and strict application of law, and negotiated compliance, that is, bureaucratic flexibility and bargaining with the regulated industry (Hunter and Waterman, 1992). Problems with the first are notorious. In British Columbia, for example, effluent permits were for years routinely exceeded by almost everyone, with absolutely no consequences. Certainly there will be problems with the second – the likelihood of secret deals between industry and regulators – but these might be expected to diminish if there were broad stakeholder participation in the enforcement process.

Thirdly, when central government agencies promulgate national environmental standards, the costs of meeting those standards – for example, for drinking water quality – often fall on regional or local governments. In the US, cities and school districts are rebelling against the high costs of meeting national standards under circumstances where they have had no direct input into the formulation of those standards, either in terms of evaluation of the scientific evidence or the determination of acceptable risk. 'Nine participating cities in Ohio have made an important, detailed study of impacts on them of 14 environmental regulations or issues. They estimate their compliance costs (1992–2001) at about $3 billion' (Abelson, 1993: p. 159). The best example is asbestos removal which impacts heavily on the budgets of welfare institutions such as public schools and hospitals. The predominant expert opinion is that where asbestos insulation is in good condition it should be left in place (D'Agostino and Wilson, 1993); these types of decision arguably are made better if a full range of local stakeholders is involved.

Finally, an argument has been made recently that preventive design – a form of environmental regulation which requires a demonstration by the initiator prior to permitting the use of a new technology or facility that everything has been done to minimize adverse environmental impacts – is far superior to the traditional strategy of simply permitting certain levels of pollution based on theoretical studies of probable harm (M'Gonigle *et al.*,

1993). This new approach would work best where stakeholders were involved in all stages of the design and implementation process.

MANAGING RISKS THROUGH NEGOTIATED CONSENSUS

There is enough experience to date in North America with negotiated rule-making and consensus-building processes to suggest that the parties will indeed come to the table (Pritzker and Dalton, 1990). Usually this is because government agencies with statutory authority behind them have indicated it would be a good idea for them to do so, although non-governmental parties can also initiate the process. Those with a direct stake in the outcome (usually industry and labour or industry and environmentalists) have little choice but to appear; those with an indirect stake have more flexible options, but they could not avoid all such events and still retain credibility as a stakeholder in risk debates. Often these are not happy events because risk controversies are 'dirty' affairs with high stakes for all the parties involved, and prior experience gives them little reason to trust each other (Leiss, 1993).

Firstly, all the risk controversy cases analysed in the literature have a common feature: at the point when the controversy erupted some risk assessment data which were regarded as indispensable from the standpoint of at least one key stakeholder were unavailable. Since this is not the kind of data which can normally be generated on the spot, this deficiency always goes unremedied for a considerable length of time. Depending upon how long the initial controversy lasts, data may still be missing when the controversy dies down again. Nevertheless, the inherent nature of face-to-face meetings, as well as the need for all parties to take reasonable positions in order to demonstrate that each is serious about reaching consensus, means that the outstanding deficiencies in the risk assessment database will be regarded as regrettable. If completing the database is considered to be essential to an agreement, a commitment by the risk promoter to do so expeditiously will almost certainly be forthcoming.

Secondly, the representatives of those stakeholders who bear the most immediate level of excess risk will normally insist upon choosing independent expert consultants: (a) to evaluate the pre-existing risk assessment (or if none is available to prepare one); (b) to present their findings in terms comprehensible to the non-expert; and (c) to engage in dialogue about what ought to be considered an acceptable level of risk under the particular circumstances at hand. Even if the initial findings were to show unacceptable levels of excess risk to a key stakeholder, this need not be more than a temporary obstacle to a settlement: some risk reduction technologies or strategies are almost always available to bring the numbers within acceptable limits. This is a matter of money, capital investment or worker training and safety equipment programmes, and thus becomes part of a straightforward risk/benefit trade-off to be made at the table by the risk promoter.

There are enormous advantages inherent in the immediacy of feedback which is available when the parties are all sitting around the table. This is bargaining in the classic sense, which will involve bluff, bluster, threats, portents of doom, denunciations, mock outrage, recriminations and the like. Those who are concerned that the level of excess risk is too high and who propose that the risk promoter commit funds so as to reduce it, take a chance that the expenditures demanded will tip the promoter's benefit–cost ratios against undertaking or continuing the project, with the concomitant threat – to labour, the local community and the regional government – of employment and monetary losses. At any point in the process this threat may be only bluff and the risk promoter indeed may be willing to put more money for risk reduction on the table later in return for concessions in other areas. The others have to evaluate their bottom line acceptable risk numbers, including the uncertainty which ranges around those numbers, taking into account the fact that *all risk estimates represent possible outcomes only and that the actual adverse health effect may be zero*. If any will object that this appears to be 'bargaining with people's lives' the rejoinder is: correct, so it is. Furthermore, every one of us, every day of our lives, willingly – albeit more or less wittingly – engages in the same sort of activity, both for ourselves, and for those dependent upon us, as we make a running series of judgements in choosing what foods to eat, how to operate on city streets, what medical and exercise regimes to follow, what discretionary pleasures to indulge in, and so on.

To summarize, these are types of specific risk debates which will characterize the dialogue between the risk promoters on the one hand, and those constituencies most directly affected by the possible elevated level of risk on the other. Examples of the two sides are: industry and labour unions for new or newly assessed occupational hazards; industry and a local community for siting a chemical manufacturing plant; government and one or more local communities for siting a waste treatment facility; government and industry, for regulations setting limits on discharges of potentially hazardous materials into air and water or just scientific standards on how to measure those discharges. Among other topics, the debates will centre on the availability or quality of the existing risk assessment, the credibility of that assessment for all stakeholders, the extraction of commitments for an improved (independent) risk assessment if necessary, the determination by all affected stakeholders of the limits of acceptable risk, and the negotiation of risk reduction strategies in the light of their impact on benefit–cost ratios.

THE BRITISH COLUMBIA ANTISAPSTAIN CASE

Antisapstains are pesticides (fungicides) used to control the growth of mould on freshly sawn softwood lumber which would otherise result in highly noticeable stains on the lumber surface which will make it difficult or impossible to sell. The events of 1989–90 described below were the result of the lumber industry's decision to discontinue their use of the chlorophenate compounds as antisapstain agents, despite the fact that they had given excellent results at a relatively low cost for about 40 years in Canada and elsewhere (Leiss and Chociolko, 1994; Leiss, 1992). The decision was the result of determined pressure from labour unions concerned about both long-term occupational health effects and environmental impacts (dioxin contamination and persistence in soil and water). Alternative registered products had been introduced in the late 1980s, but the most common of these also gave rise to strenuous union protests, based on, in this case, acute health effects (skin irritation) and a call for its elimination as an option in the mills. The industry replied that no other efficacious registered products were available for use; the Federal regulatory agency, Agriculture Canada, added that no new antisapstain products were likely to be registered for quite a number of years because of serious gaps in the required toxicology databases. At the same time, the British Columbia provincial government was warning that existing uses probably would not meet its new effluent regulations for allowable concentrations of chemicals in stormwater runoff from the mill sites.

Thus, there was a standoff with a highly uncertain outcome for an industry of great economic importance to Canada, which had, as of 1989, about 40 per cent of the world's softwood lumber market. The issue was of particular importance in British Columbia which is a major producer within Canada. Many of the parties had met continuously for more than 5 years to find a way out of the impasse which had been developing slowly, to no avail. In the meantime, the exchanges had grown increasingly bitter. Most expected Agriculture Canada to take some action which would resolve the matter. However, when that agency stated firmly in late 1989 that it saw no way of doing so, the parties realized that they had no alternative but to explore the one option which the agency was prepared to leave open: the British Columbia stakeholders should meet together to see if they could agree to support some solution to the impasse, including a recommendation that the agency should register one or more chemicals which did not have adequate databases. If they did so, Agriculture Canada would 'probably' take regulatory action consistent with those recommendations. In the first half of 1990, 11 organizations sent representatives to the table: four labour unions, three industry sectors, two provincial government departments, and two environmental organizations (the Federal agencies maintained an 'arm's-length' stance so as to keep their own options open).

Of all the people in the room the labour union representatives faced by far the most agonizing choices, for the real consequences of the risk/benefit trade-offs being discussed fell squarely on their shoulders. The dilemmas which they faced related to both health and economic risks. There were three dimensions to the health risk:

1. giving explicit consent to the use of a new chemical with an incomplete toxicology database, the use of which might later be proved damaging to the health of the people they represent, was something that had not been done before. What was the probability of this outcome?
2. If there was no agreement on a new chemical, Agriculture Canada's *status quo* option meant that the existing chemicals, to which their members objected and at least one of which had known adverse acute effects, would continue to be used. What was the probability that this was a worse option than agreeing to a new chemical?
3. What was the probability that the industry was serious about the option of reinstating use of chlorophenates – about which there were fears of long-term chronic adverse effects, especially excess cancer mortality?

At the same time they had to consider the adverse economic consequences to their members of a failure to find any solution, namely the risk to their jobs. What was the probability, first of all, that there really was no viable non-chemical technology and secondly, that a failure to find any solution could result in significant job losses?

In the end, union representatives decided to accept the responsibility for answering these questions for themselves – in other words, for making judgements on acceptable risk and risk/benefit trade-offs which formerly would have been made by regulatory officials.

After some difficult meetings an agreement was signed by a majority of the parties which was accepted by Agriculture Canada as a stakeholder consensus and the Federal agency implemented all recommended actions in the agreement. The basis for the agreement was the acceptance by other parties of the unions' demand for an effective ongoing voice in the micro-management of risks and benefits in their industry. This was accepted by the regulatory authorities, who have made this stakeholder group a permanent part of the process for regulating antisapstain use in Canada. This ongoing activity takes place in two ways: (a) through a joint labour and industry health monitoring committee which evaluates chemical use practices at the mill sites; and (b) through the indefinite continuation of the all-party multi-stakeholder meetings, where all new regulatory developments are discussed before implementation and where chemical industry representatives may present arguments in favour of new products which promise to reduce both occupation and environmental risks.

CONCLUSION

Controversies over setting and enforcing standards for managing health and environmental risks have many seemingly intractable features which defy easy resolution. When risk management decisions are unpacked, they reveal themselves to be made up of myriad qualitative judgements, in all of which competing social interests and community values play a part. Moreover, they are never settled once and for all: new knowledge, new technologies, new values and new voices constantly change the bases of judgement in these matters. Most importantly, the bottom line determinants of acceptable risks and acceptable trade-offs are by their very nature questions of fundamental values, which in a democratic society ought to be regularly referred to representative groups of citizens for their consideration. In this light, the tasks of setting and enforcing environmental standards are best conceived as a process of continuous micro-management which requires the participation of a broad range of stakeholders.

6. Problems and issues for public sector involvement in voluntary standardization

Jacques Repussard

One of the best descriptions of the objectives of voluntary standardization is as follows :

> Standardization is the voluntary and methodical harmonization of material and non-material objects undertaken jointly by the interests concerned for the benefit of the community as a whole. It shall not lead to individual interest gaining a special economic advantage and requires consensus agreement between all parties concerned.
>
> The standards cover products, systems, processes and services and promote rationalization and quality assurance in the fields of industry, technology, science and administration. They ensure interworking and interchangeability of products and systems, rational order and communication between market partners. They are aimed at supporting the free movement of goods and services within Europe, at protecting the safety of persons and of property and at providing an improvement in the environment and in quality in all fields of life. (CEN/CENELEC/ETSI, n.d.).

From this description it can be readily seen that government authorities often have a direct interest in the operation of standardization.

ASPECTS OF GOVERNMENT INVOLVEMENT IN STANDARDIZATION

Concern for the efficiency of the country's economy leads governments to take a general interest in standardization. Governments may wish to promote standards for the development of technical infrastructures. Standards can also play a role in economic restructuring. Following the Second World War, for example, some countries linked reconstruction programmes to tightly controlled standardization measures. Governments may also have an interest in promoting and facilitating better access to the standardization process for small and medium-sized enterprises, consumers, trade unions, research institutions and so on.

Governments are also aware that, in some ways, standards-making is an activity which derogates from the usual antitrust rules and, as a result, strict rules for guaranteeing openness of the standardization process have to be enforced. Finally, standards have a direct connection with international trade, since they can be used as trade barriers. This has led the General Agreement on Tariffs and Trade (GATT), as well as the European Community (EC) and the European Free Trade Association, to develop policies aimed at the elimination of the use of standards as barriers to trade.

A government must also be concerned with the protection of its citizens through safety, health and environmental standards. Recognizing the '*quasi* public-service' nature of the standardization process, some governments have been inclined to rely on it as a support for regulatory functions. This policy is known as 'reference to standards', and is a feature of regulatory systems in the EC.

Finally, governments must oversee the operation of the public sector. According to each country's traditions and economic history, its public sector can be wide or narrow. At the very minimum, administrations must manage matters of 'public procurement'. Governments have often used standards as a tool to rationalize this process.

RELATED PROBLEMS AND ISSUES

The Principle of Public Sector Involvement

In observing what is commonly accepted to be 'voluntary standardization' in different countries of the world, an immense variety of organizational situations becomes visible. These vary from totally private institutions with virtually no central government influence or involvement, to those totally run by the state, and operating according to state controlled procedures.

Standardization has to be looked at as a reflection of the organization of society in a given country. In this light, we must consider whether the status of standardization is coherent with the overall structural organization of a society at a given moment in time. This is why, for example, it can be dangerous to promote in eastern European countries a particular model which has proven successful in a different kind of society. Standards are best set and used in the context of a national system that is well integrated into the mainstream organization of a society and economy. However, because the mainstream of economic policies accept and promote the development of international trade, fewer and fewer standards are being set at national levels.

This is why the European Committee for Standardization (CEN) is developing standards at the European level, but is also committed to the principle of implementing European standards according to existing national social and economic conditions.

Public Sector Involvement in 'Upstream' Standardization

Governments may become involved at the standard setting level and can influence the process through administrative controls and financial incentives, and also by indirect means.

Administrative controls may include regulating the process itself, as, for example, through a 'standardization law'. They may also involve giving 'public' status to the standards body, its procedures and to the standards it produces. One result is that conflicts over standards can often be dealt with in administrative courts rather than civil ones.

Administrative controls, including granting public status to the standards body and heavy public financing, may in some cases yield the result that the economic partners in standardization do not take sufficient responsibility for setting priorities and programming the work. On the other hand, in countries where there is a strong tradition of powerful administrative intervention, this type of control may be necessary to establish the very credibility of the standards.

In the case of CEN, it is to be noted that administrative control (at the European level) has been limited to the negotiation, in exchange for granting of status (EC, 1983), of 'policy guidelines' that encompass the key principles of democracy and openness to which CEN has pledged to operate.

Financial involvement, where it exists, takes the form of public subventions, or of contracts linked to specific services provided by the standards body (for example, the elaboration of some standards). The problem with subventions is that their relative level is sometimes difficult to justify. There is little precise information on the economic benefits of standards, or on the costs of non-standardization. Furthermore, subventions usually carry with them a notion of overall relative control by the subsidizing authority, and often tend to be revised for reasons linked to overall budgeting constraints rather than standards-related issues. Contrary to accepted belief, therefore, heavy subsidizing may be an element of instability in the medium term, which can be critical since standards-setting is by nature a medium-term operation involving 3–5-year-programmes.

Contracts, on the other hand, presuppose that the cost of the service has been evaluated. Again, there is a relative lack of precision in the available data concerning the cost of standards-setting. Figures are relatively well known for central administration costs, but these are not the most important ones. Data on the financial aspects of standardization are 'unofficial'. In most organizations, programming and financing are dealt with separately and the adjustment factor is the time needed to implement the programme. Since the real economic benefits to be gained from the use of a future standard are not usually quantified, uncertainty over the time factor is in turn not perceived as a serious problem (which is one of the reasons why many CEN programmes sponsored by public budgets are running late).

Governments can apply many forms of indirect influence, and it would be useful to conduct case studies to evaluate the efficiency of such approaches. Among the most common are the following:

- participation of government experts in the standards drafting groups;
- support (usually financial) for particular groups for their participation in standards drafting (consumer groups, trade unions, small or medium-sized enterprises);
- education and promotion;
- allocation of research and development funds to help laboratories prepare technical input into standardization;
- reference to standards in technical legislation, along with the setting of specific requirements for such standards.

Public Sector Involvement in 'Downstream' Standardization

Public policies also have impacts on the relative level of use of published standards, and upon their effects. The question concerns the extent to which public sector involvement leads to a situation where the standard loses its original voluntary status.

A typical example of this problem is the well-known ISO 9000 series on Quality Assurance. These standards were originally designed in order to 'standardize' the pattern of Quality Assurance requirements from contractors to their suppliers, thereby simplifying the process of dealing with several contractors, and contributing to better overall quality management.

However, should these complex standards start to be systematically applied (requiring certification) as a prerequisite for doing business, it is obvious that the original goal of the standard has been forgotten, and that the overall economic cost of operating the standard could become higher than the resulting benefits. This scenario could be encouraged through government promotion, specification of standards in public procurement procedures or even through legislation. The logic that, once defined, standards yield optimal economic benefits when applied across the board, may not apply systematically to systems standards that cover concepts like quality, safety and the environment.

Other downstream problems concern such matters as discrepancies of knowledge about standards within public administrations, the diffusion of standards themselves in ministries and so on, and, eventually, the monitoring of jurisprudence in conflicts involving standards issues.

In conclusion, it can be seen that government authorities have at their disposal many tools with which they can play a role, upstream as well as downstream, in the evolution of standardization policies. The type of tools which are actually used, and the goals of governmental interventions, vary considerably at a given moment depending on the overall context of the economic and regulatory policies in different countries or regions.

The present generally accepted international trend towards liberalization of economic exchanges and deregulation does not mean, in this context, that governments should be, in principle, less concerned about standardization. On the contrary, the present period should see more focus on avoidance of technical barriers to trade, and on the benefits that can be gained from voluntary standards forming a 'level playing field' for the economic players, in a context of growing deregulation.

7. International challenges in defining the public and private interest in standards

Stanley I. Warshaw and Mary H. Saunders

We are on the edge of a revolution in the development of internationally recognized standards and conformity assessment systems. Increasing global trade and technological integration raise important issues concerning: (a) existing infrastructures and their efficacy for developing internationally accepted standards; (b) the interaction of conformity assessment activities and systems among market-led economies; and (c) the changing prerogatives of governments. Our changing global economy demands that both government and business pay more attention to domestic and international standards-related issues and activities.

As a basis for world trade, manufacturers and processors are demanding single globally acceptable technical standards and conformance tests. However, national governments must accommodate domestic health, safety and environmental goals that often differ dramatically between countries and regions. How then can we ever achieve the development and implementation of global standards and the acceptance of conformity assessment measures worldwide?

STANDARDIZATION IN THE US AND NORTH AMERICA

The role of the US in the world economy is changing. Historically, US business could rely on the fact that US technology, standards and technical specifications could dominate most domestic and foreign markets. This is no longer true and economies of scale are now measured in terms of global, not national, markets. Incompatible national and regional standards raise the cost of doing business globally and more effective participation by US interests in international standards activities has become an economic imperative.

The Congress of the US has conducted a number of hearings in the past few years to determine what might be accomplished through legislation to improve the international effectiveness of the nation's standards infrastruc-

67

ture. Congress has also commissioned several studies of this issue. Agencies such as the Department of Commerce and the National Institute of Standards and Technology (NIST) have also conducted public hearings and put forth proposals to facilitate the harmonization of national and international standards and conformity assessment measures.

For a number of reasons, the government's stake in standards-setting will loom even larger in the future. The hundreds of standards development bodies in the US are becoming increasingly competitive. There is also a dichotomy between standards developers representing professional societies and those representing trade associations. The number of private sector organizations involved in standards development in the US is increasing – to more than 600 at a recent count.

This decentralized system of standards development has led to the perception among the UK's trading partners that the US lacks a single voice in international fora. Differences in viewpoints and objectives among major private sector standards-development organizations in the US have made it sometimes difficult to arrive at a unified position. These organizations have also found it difficult to reach consensus on the appropriate role to be played by government.

The US government has been engaged simultaneously in negotiating the North American Free Trade Agreement (NAFTA), and the Uruguay Round of General Agreement on Tariffs and Trade (GATT). Standards issues are key elements of both. Under the GATT Agreement on Technical Barriers to Trade – the 'Standards Code' – signatories must use international standards as a basis for technical regulations or standards, except where these are inappropriate for the protection of the public or national security interest. The Code also commits signatories to participation in international standards activities that support areas of regulatory interest. The Code does not address direct participation in the development of national standards, however. There are no requirements in the existing Code, or in revisions contemplated under the Uruguay Round, that open the domestic standards development process to outsiders. Nevertheless, most standards-developing entities in the US are open to foreign participation, particularly those operated by professional societies which, incidentally, develop most of the voluntary standards in the US.

NAFTA encourages the use of international standards as a basis for regulation. International standards are those adopted by an international standardizing body which is open to the standardization bodies of the signatory nations to the GATT Code. NAFTA also requires each party to the agreement to recognize conformity assessment bodies in the territory of the other parties. The principles of openness and broad participation by foreign and domestic parties characterize both US standards development activities and the standards-related activities planned under NAFTA.

STANDARDIZATION IN THE EUROPEAN COMMUNITY –
AN ALTERNATIVE MODEL?

In contrast to the US system, standardization in the European Community (EC) has significant government influence. The EC system is much more integrated at both the national and regional levels. It is also generally a closed system. Participation in formal standards development activities at the regional level is limited to authorized representatives of European national standards bodies, thus excluding participation by non-European countries, even the signatories to the GATT Code.

European regional standards bodies – European Committee for Standardization (CEN), European Committee for Electrotechnical Standardization (CENELEC) and European Telecommunication Standards Institute (ETSI) – have been contracted by the Commission of the European Communities to develop standards in support of EC legislation. These include standards to support legislation for product safety, workplace and environmental regulation, and EC procurement. Regional standards in Europe preempt national standards, and are given preferential status under EC legislation. Products that conform to European standards are presumed to meet EC essential requirements, as specified in directives.

The regional harmonization of European national standards will have a growing influence on the development of international standards. In negotiations with the Commission of the European Communities – detailed in a June 1991 Joint Communiqué between US Secretary of Commerce Robert Mosbacher and European Commission Vice-President Martin Bangemann – the US government has been unsuccessful in gaining 'a seat at the table' in European regional standards bodies although a commitment was obtained in 1991 for the EC to give preference to the adoption of international standards before undertaking a regional standards development activity, as long as the international standard met EC needs and time constraints.

CEN and CENELEC responded to the US government request for openness and transparency by instituting a formal policy regarding relations with non-member countries. It provides for proposals and comments to be made by non-European interests through their national International Organization for Standardization (ISO)/International Electrotechnical Commission (IEC) member bodies. Additionally, CEN and CENELEC have effected agreements with ISO/IEC, that allow for coordination of standards activities and parallel voting on draft standards, ensuring the technical equivalency of international and European regional standards.

This access falls far short of direct participation in European standards activities. While third countries, such as the US or Japan, can influence the development of an international standard through their participation in the relevant international technical committee, they cannot contribute to the important deliberative efforts during the drafting of European standards. This increasingly becomes an issue as the current Western European members of Europe's regional standards bodies extend observer status and possibly full membership to selected Eastern European nations. ETSI, the only European regional standards body which is open to direct participation by non-European companies, through associate membership, has recently adopted a discriminatory Intellectual Property Rights (IPR) policy which may force major non-European IPR holders to withdraw from ETSI membership.

INTERNATIONAL STANDARDS – THE US OR EC MODEL?

Which system for international standards development can best meet the changing needs of global businesses? One which resembles the European regional system – monolithic, integrated, formalistic and policy-driven; or a system more like that which has evolved in the US – pluralistic, sometimes fragmented, *ad hoc* and market-driven?

Superficially, the international standards system, as represented in ISO/IEC looks more like the European model. However, international standardization activities extend well beyond ISO/IEC, and there is real competition among international standards bodies in several areas. As in the US system, competing organizations seek to respond to changing priorities and interests.

Three areas of standardization illustrate this point. In the information technology area, both the United Nations Economic Commission for Europe (UNECE) Working Party 4 and ISO/IEC's JTC1 are engaged in developing standards for Electronic Data Interchange (EDI). As UNECE Working Party 4 activities expanded beyond Europe, participants voted to change the structure of the group to reflect its international activities, and to coordinate UNECE and JTC1 activities.

On the other hand, UNECE activities have conflicted with those of ISO in the transportation area. Both the UNECE Working Party 29 (WP 29) and various ISO committees set standards for motor vehicle equipment and parts, road safety and pollution and energy controls. In 1989, WP 29 issued a mission statement noting the new goal of promoting world-wide harmonization of motor vehicle regulations. However, the group's proposed new rules of operation to support this goal do not adequately address the issue of coordination with other international bodies (such as ISO) and would perpetuate a European-dominated voting structure.

In the environmental area, both ISO and the Organization for Economic Cooperation and Development (OECD) have undertaken standards-related activities. Moreover, the Business Council for Sustainable Development, an international group of business and industry executives, made a proposal for the development of international standards for environmental performance to the UN Conference on Environment and Development.

International standards bodies must show increasing flexibility in meeting emerging requirements to support international business. Some organizations will have to take on new standards activities, broaden participation and/or change their approach to standards development to meet these requirements. The US model, which provides for broad participation of all interested parties and which allows competition among parties to determine the best technical approach may be the best one to enable international standards bodies to respond to the growing demands placed on them.

CONFORMITY ASSESSMENT AND TRADE

To sell in world markets, today's manufacturers face increasing demands for independent third party assessment of the conformity of their products/processes to specific standards and/or legal requirements. These demands go far beyond those areas historically regulated by governments. New requirements are arising for quality systems registration (ISO 9000), and for certification of the environmental characteristics of certain products and processes. Many developing countries maintain broad certification requirements covering safety of consumer goods and industrial products imported into their territories. Although enforcement historically has been poor, countries such as China and the Russian Federation are now stepping up their certification programmes.

Uruguay Round revisions to the text of the GATT Standards Code recognize conformity assessment as a potential trade barrier. The text requires nondiscriminatory application of conformity assessment programmes by central government bodies. It specifies the use of international guidelines and recommendations and establishes a dispute settlement procedure. This text encourages signatories to accept foreign conformity assessment bodies in their systems under the same conditions as bodies located in their territory (or the territory of any other country). However, it also recognizes the need for signatories to negotiate mutual recognition agreements.

The US and EC approaches to conformity assessment provide a study in contrasts. As part of its internal market integration programme, the Commission of the European Communities issued a policy document in 1990 on conformity assessment – the 'Global Approach to Certification and Testing'. The EC system relies heavily on product and process assessment by independent third parties – product testing, type examination, quality system certification and production surveillance, where appropriate. The system permits authorization of entities located in a third country to perform third party certification, accreditations or approvals, provided that these entities operate under a mutual recognition agreement negotiated between the Commission and government authorities of that country.

In contrast, US conformity assessment activities are largely decentralized and market-driven. There are relatively few programmes requiring product certification at the federal level – many of these programmes recognize manufacturer self-certification or certification by foreign laboratories. Some certification programmes are administered by state or local governments in the US. Finally, there are voluntary private sector certification programmes that are not mandatory, but respond to consumer demand. In some cases, these are a voluntary industry response intended to forestall action by a regulatory agency. These programmes co-exist with government-adminis-tered programmes to meet regulatory goals.

While the US does not have a 'national' accreditation system, there are programmes at the federal government level that can assure the technical competence of participating US entities. The National Voluntary Laboratory Accreditation programme, operated by NIST, provides unbiased third party evaluation and recognition of laboratory performance, based on international guidelines. The programme is designed to be compatible with domestic and foreign laboratory accreditation programmes.

NIST has also established a National Voluntary Conformity Assessment Evaluation programme, designed to provide US government recognition to US-based conformity assessment bodies which test or certify products for acceptance into foreign *regulated* market areas. The programme will provide accreditation or 'recognition' in the areas of product certification, laboratory testing and quality system registration in conformance with international guidelines.

INTERNATIONAL ACCEPTANCE OF CONFORMITY ASSESSMENT MEASURES

As noted above, both the US and the EC have developed structures to support mutual recognition of conformity assessment activities outside their territory. Negotiations in this area have proceeded largely on a bilateral basis – a pains-taking and often repetitive process. For example, the Commission of the Euro-pean Communities has indicated that it cannot support bilateral mutual recog-nition negotiations initially with more than three countries. Those that do not make the 'first cut' may have to wait several years for mutual recognition.

There are some existing models for broader international conformity assess-ment systems which provide for recognition of conformity assessment results across multiple markets. These include the certification programme for measuring instruments operated by the International Organization of Legal Metrology, and two IEC programmes – the IEC Quality Assessment System for Electronic Components and the System for Conformity Testing to Standards for Safety of Electrical Equipment. These international certification programmes have met with varying degrees of market acceptance. In some cases, certification requirements were felt to be too cumbersome. In other cases, there was competition from similar regional programmes. These programmes are also weakened by the fact that in no case does certification substitute for the requirement to obtain national or regional approvals to meet regulatory demands.

While there are international conformity assessment programmes operating in specific sectors, such as the electrotechnical area, there is no broad recognition programme covering a variety of sectors. The lack of international recognition is keenly felt in the quality system arena, where ISO 9000 certification to meet foreign market and regulatory requirements may mean multiple registrations for a company, with no value added. If we can reach broad agreement on mutual recognition of accreditation programmes operated as prescribed by international guidelines, then certification activities undertaken in any world location by accredited entities could be accepted by participating governments without further negotiation being necessary.

CHANGING GOVERNMENT PREROGATIVES – THE US EXPERIENCE

The link between standards and regulation is clear across all sectors where health, safety and environmental interests are at stake. However, this does not mean that every standard should become a regulation. In the US we have taken the approach that regulatory action is taken only in those instances where the market has demonstrated that it is not capable of self-regulation. Regulations observe national and sometimes regional boundaries (as in the European Community), whereas markets increasingly do not.

In the US and other industrialized countries, we are seeing a significant change in the perspective and outlook of regulators regarding international standards. There is a growing awareness of the need to address many regulatory issues on a regional or global basis. Some US government agencies have made a policy commitment to become more involved in the international standards arena, as a means to this end. Standards and regulatory convergence have also become critical elements of bilateral and multilateral trade negotiations.

US regulatory agencies at both the federal and local levels have also become increasingly sensitive to the international impact of testing and certification requirements. This is in part because more foreign products are entering the US marketplace. However, there is also a growing sensitivity within the government about the potential impact of domestic regulatory requirements on the competitiveness of US manufacturers in foreign markets.

Agencies such as the Food and Drug Administration (FDA), NIST, the Federal Communications Commission and the National Institute of Occupational Safety and Health have initiated inquiries or rule-making procedures to develop the basis for harmonizing their regulatory requirements with the international marketplace. Representatives of FDA and the Environmental Protection Agency are engaged in discussions with their European Community counterparts regarding mutual recognition of Good Laboratory Practice audits for chemicals and pharmaceuticals, based on OECD guidelines. US agencies are increasingly relying on voluntary standards as the technical basis for regulations and are looking to international quality systems standards to help reduce procurement and administration costs. Finally, federal agencies are taking the lead in developing accreditation programmes that meet international guidelines and are also supportive of mutual recognition of conformity assessment.

CONCLUSION

Both the US federal government and the US private sector are more supportive now of global standards and mutual acceptance of conformity assessment measures world-wide than at any previous period. The incentive for private sector support and interest is the potential for reducing the cost of doing business globally.

There are clear parallels in other countries as well and, thus, there may now be a good basis for achieving international harmonization and mutual recognition in standards-related areas. Such initiatives have the greatest chance of success in areas where clear economic benefits can be identified by industry, and where a cooperative relationship can be established between regulators and the industries they oversee to ensure that national environmental, health and safety goals are not impaired.

Harmonization efforts, such as those considered between the US and the EC in areas such as the environment, food safety and aircraft safety may also facilitate parallel initiatives to develop global standards and conformity assessment programmes. The challenges in international standardization that face all nations can be met only through cooperative efforts addressing market needs. The resulting infrastructure, monolithic or pluralistic, must be responsive to the marketplace, and more especially, to the varied and changing needs of different industrial sectors and of governments.

8. Process environmental audits and the proposed British standard on environmental management systems

Anthony K. Barbour

INTRODUCTION

During 1993, environmental management systems were the subject of wide-spread activity. Formal proposals included British Standard (BS) 7750 in the UK and the Eco-Management and Auditing Regulation (EMAS) in the European Community (EC) as well as other initiatives in France and South Africa. Both BS7750 and EMAS are basically concerned with management systems which will ultimately be capable of external accreditation by qualified verifiers. They are not directly concerned with the measurement of environmental performance, though EMAS places more stress on this aspect than does BS7750. The proposals outline management systems which, if correctly resourced and fully implemented, will assist the identification of environmental issues and problems and thus, ultimately, improve performance.

This Chapter shows how environmental reviews and auditing standards can assist both operating and main-board managements to achieve and maintain the high level of environmental, health and safety performance now expected from the process industries. In this endeavour, environmental audits complement other management tools such as quality assurance and risk management. BS7750 and its relationship to EMAS are described before moving on to consider the scope, objectives, execution and implementation of site environmental audits. In conclusion an assessment of the role of BS7750 and issues relating to its implementation are presented.

BS7750 AND EMAS

BS7750

BS7750 will shortly be published in its final form and will take into account a wider consultation exercise and the comments of over 40 working parties spread across virtually the whole of UK industry. Over 400 companies participated actively in the pilot phase and the final product should be acceptable to the whole range of UK industry, not only sites and processes which are potential major sources of pollution and subject to regulation by Her Majesty's Inspectorate of Pollution (HMIP) and the National Rivers Authority (NRA). The UK mechanism for accredited certification is currently subject to consultation by the Department of the Environment (DoE) and the Department of Trade and Industry (DTI) and has not yet been promulgated.

Relationship with EMAS

Auditing is only part of BS7750. In its final form, the standard will be compatible with the EC EMAS proposal. The final EMAS draft included a requirement that emissions should be compatible with the principle of 'best available techniques not entailing excessive cost' built into the EC's 1984 air framework Directive and the Integrated Pollution Control (IPC) regime established under the UK's 1990 Environmental Protection Act. Whilst this requirement, introduced at a very late stage, is acceptable to companies already regulated by HMIP, it is largely a new requirement for others. Ingenuity will be required to ensure that the final wording will be acceptable to all companies wishing to adopt BS7750. Table 8.1 summarizes the essential objectives of the UK and EC schemes.

   If the two schemes are to be made fully compatible, accredited certification is necessary. This will require a clear understanding between companies and certifiers regarding the scope of the standard, bearing in mind that BS7750 is about environmental management systems and not specifically about environmental performance. Superficially competent systems which fail to deliver effective environmental performance could be a problem for both management and certifiers.

Organization for Implementation

Implementation of an effective environmental management system certificated under BS7750 will require the acceptance of a company (and, where appropriate, a site) environmental policy which is publicly available, sets clear environmental objectives and demonstrates commitment to continual

Table 8.1 Comparison of BS7750 and the eco-management and audit scheme

BS7750	EC eco-audit/management
Environmental policy oriented	Objective oriented
Relevant to activities, products and services, and their environmental effects	Establishment and implementation of environmental policies, programmes and management systems by companies, in relation to their sites
Understood, implemented and maintained at all levels in the organization	
Publicly available	Systematic, objective and periodic evaluation of the performance of such elements
Commitment to continual improvement of environmental performance	Provision of information on environmental performance to the public
Sets and publishes environmental objectives	Later, a commitment towards continuous improvement of environmental performance

environmental improvement. Furthermore, implementation will require industry to be committed to effective training, continuous updating of information on the environmental effects of its processes and products, detailed knowledge of the laws and regulations under which it operates, regular environmental auditing based on detailed record-keeping, and continual review of the implications of environmental audits.

PROCESS ENVIRONMENTAL AUDITS

Company managements at the highest level are increasingly accepting responsibility for setting the framework through which operating plants and the products which they manufacture meet, on a continuing basis, all relevant environmental and safety standards. They must also ensure that sufficient funds are available to finance the capital expenditures necessary to implement environmental policy as it evolves in detail at operating plant level.

High-quality environmental performance within a company is far too important to be left solely to those with the label 'environmental' or 'health and safety' in their job title. High performance must be completely integrated into the company ethos. This will be achieved only if the practical issues which it raises receive serious analysis and commitment from every department – not only production and public relations but, vitally, engineering, finance, legal, sales and research.

The Scope of an Audit

Environmental and safety standards may be expressed through either quantitative or subjective criteria. Quantitative criteria may be applied to atmospheric emissions, liquid effluents, the purity of the in-plant atmosphere, biological parameters for some industrially exposed workers, permitted impurity levels in drinking water and some foods, water and air quality standards, and product safety.

Subjective issues, which form an increasingly important component of company environmental policies, may cover externally perceived odours, criteria for solid waste disposal, sources of raw materials, major accident hazards and their prevention, recycling, environmental acceptability of products and eco-labelling, and methods of testing for product safety.

In addition, several global environmental issues could have major impacts on the process industries even though the problems themselves cannot be quantified accurately. Examples are the forthcoming ban on the use of 'hard' chlorofluorocarbons to arrest further deterioration of the stratospheric ozone layer and attempts to reduce the felling of timber in the tropical jungle.

Many environmental issues in the process industries, particularly those for which quantitative criteria can be applied, must be identified and solved at the operating plant level. However, environmental concerns are now felt by people who live close to plants or to repositories for unsaleable wastes, by consumers of products nationally and internationally, by providers of raw materials and by those who may judge that activities are irretrievably disrupting the ecological balance of the world. Thus, environmental audits in the process industries should analyse both quantitative and qualitative issues.

The Objectives of Environmental Audits

In the process industries, the objectives of an environmental audit are: (a) to assess compliance with all relevant current environmental, health and safety criteria and, if necessary, to develop programmes for improvement in conjunction with plant management; (b) to advise operating management of likely future regulatory criteria and to develop systems and necessary capital expenditure programmes to meet these future criteria; (c) to ensure that all product safety regulations are fulfilled and to review the health and environmental acceptability of the product line, with appropriate recommendations;

and (d) to reduce as far as possible the major accident hazard potential of operating plants.

With the rapidly increasing awareness of environmental, health and safety issues and the enormously increasing cost of both physical remediation and human health claims, it is now commonplace for one company wishing to acquire another to conduct environmental, health and safety audits. This process is often termed 'due diligence', following US practice. An environmental, health and safety audit is part of the overall legal and financial review. In this case, the principles and objectives of the audit are the same, though the timescale is often shorter and the familiarity with the plant, its management and their records is much less.

The overall objective of auditing is to assist management at all levels to ensure routine compliance with current and future regulatory criteria. In relation to technical issues, the objectives of the auditor and management are the same. However, senior boards and chief executives are often not particularly interested in a good routine compliance record – this is taken for granted. Many senior board members are primarily interested in the avoidance of major incidents which will attract adverse publicity or court proceedings.

The objectives of auditing are mainly concerned with technical and operational issues; subjective 'political' judgements are generally restricted to estimations of when future environmental, health and safety regulatory criteria will apply to specific plants and products. The objectives are by no means all-embracing and should probably not attempt to be, otherwise accuracy and incisiveness will suffer. The main board will have major interests in many other environmentally related issues, such as capital costs, national and international competitive issues, the quality of general management, company image, public relations and the consistency of promotional literature and product safety regulations.

The most important of the many roles to be played by the main board is that of establishing and effectively promulgating the environmental framework within which the company is to operate and to provide the finance to implement environmental policy effectively.

EXECUTION OF ENVIRONMENTAL AUDITS

Preparation for the Audit

Whilst precise and detailed knowledge of the operation under review is unnecessary and may even be counterproductive, a general knowledge of the type of plant and its operational and maintenance aspects is very important.

However, the auditor must be familiar with: (a) all current regulatory criteria applicable to processes and/or products; (b) foreseen changes in these criteria and possible or likely time-scales for application; and (c) general 'environmental thinking' in relation to the products, particularly their toxicology and ecotoxicology.

For UK plants such changes are increasingly likely to be initiated by the EC which is, in turn, strongly influenced by environmental policies in Germany, the Netherlands, Scandinavia and the US. The auditor should keep abreast of all these sources of new regulation by diligent perusal of environmentally oriented journals and, equally important, participating actively in the professional discussions which occur through most trade associations. The auditor must also be aware of the known views of environmental pressure groups in relation to the plant or process and must have made a judgement of the influence which these views will have on the development of regulations.

A useful environmental audit should be an independent and knowledgeable analysis of accurate data. This requires the rapid establishment of a high level of mutual confidence – hence cooperation – between the auditor and the plant management. Audits carried out on an 'arm's-length' or confrontational basis are much less likely to be valuable in analysing the significance of issues causing plant management real concern. Information collection should take place through detailed discussion of recorded data and correspondence rather than by ticks on a clipboard questionnaire.

Content of the Audit

Professionally run plants will usually be able to demonstrate a good level of compliance measured against quantitative regulatory criteria. The main issues concerning plant managers, other than future regulatory criteria, usually relate to plant operations and latent external issues.

One of the most important plant-related issues relates to continuity and 'down-time'. Poor continuity is costly and start-ups are often accompanied by temporary environmental infractions. An audit of such situations can provoke a detailed discussion of plant operation. This should be encouraged within bounds because good standards of operation and continuity are essential prerequisites of good environmental performance. Changes in production rate, raw materials or end products can be imposed on plant management and it is vital that all likely consequences for plant operations are considered before making a final judgement. The financial benefits of greater throughput or cheaper raw materials are usually clear but the environmental consequences may be adverse. Sudden concern by the work-force about a specific health issue may also become a problem.

Progressive managements require their capital expenditure programmes to be geared to meeting foreseen future regulatory requirements as well as enabling operations to meet current criteria comfortably. The first requirement necessitates continuing familiarity with regulatory discussions in national and international environmental fora. Judging the timing of such criteria modifications is often difficult, particularly as these often originate in protracted EC discussions followed by an uncertain implementation delay.

The purification of air emissions and liquid effluents usually generates solid wastes which have to be disposed of to land in an environmentally acceptable manner. In the context of the Environmental Protection Act it is even more necessary for plant managements to visit and scrutinize carefully the standard of operation of such landfills and the means of transportation of wastes to them. Follow-up visits should be made annually in view of public and regulatory concerns which will inevitably follow the identification of particular company wastes with pollution incidents. Such pressures are, of course, encountered in a particularly acute form in the US under the Comprehensive Environmental Compensation and Liability Act which is designed to provide a funding mechanism for the remediation of defunct hazardous waste sites, but often appears injudicious in its application.

Similar uncertainties surround accurate assessment of the impact of new toxicological data and thinking on the standards of in-plant hygiene necessary to protect the health of the plant work-force. Such toxicological data and thinking will also be the major influence on product safety. Thus, from both standpoints, it is essential for the environmental, health and safety auditor to be fully aware of developments in these areas through following the advice of medical and occupational health specialists.

In the UK, in-plant hygiene is governed by the Control of Substances Hazardous to Health regulations, the application of which is now slowly maturing. Whilst the actual assessments required under these regulations are work-oriented rather than substance-oriented, an accurate knowledge of recent toxicological information is essential. Such knowledge must include an awareness of current regulatory thinking on the carcinogenic, mutagenic and teratogenic potentials of the substances being processed. Although definitive regulatory judgements often take many years to emerge, the consequences can clearly be great for the internal operating regimes of plants and product markets. Obvious recent examples are asbestos, benzene, lead, cadmium, vinyl chloride and chlorinated solvents.

Particularly in highly litigious countries such as the US, court proceedings are becoming increasingly common for compensation for alleged adverse health effects both on workers engaged in manufacturing and users of the products concerned. Such issues are thus becoming particularly important features of environmental, health and safety audits carried out as part of due diligence investigations when companies are to be acquired. In the US, millions of dollars are potentially involved if environmental or health claims are substantiated.

Whilst 'ultimate' decisions to continue or discontinue company product lines are usually taken at the highest management level, some discussion of the security of the current product line in the light of environmental, health and safety thinking is a fully justified part of the audit. Ever more complex labelling and handling requirements must be fully met; material safety data sheets or their legal equivalent must be available and in use; and marketing literature must be fully consonant with the data sheets and the requirements of the Strict Product Liability regulations. In some cases, plant may be capable of cost-effective modification to produce a line of products not subject to environmental or health pressure, in which case the audit may initiate thinking which will ultimately generate appropriate capital expenditure proposals.

At the operating plant level, a subjective judgement of good housekeeping and plant tidiness is important. Through the eyes of an experienced auditor familiar with operating – and non-operating – plants, general visible standards of tidiness are often a good guide to the quality and attitude of the management and personnel involved, and hence the likelihood of environmental infractions. Any deficiencies in these areas should be pointed out during the audit and commented upon generally in the written version. However, too much 'nitpicking' detail on minor issues which are easy to correct can cause irritation and deflect the report and its audience from the major issues.

The current climate of environmental interest and concern has initiated attitude changes in both regulatory officials and in persons living near to operating plants. Plant managements can no longer expect sympathetic treatment from either officials or local residents because of the employment the plant provides. The 'not in my back yard' (NIMBY) attitude may well dominate if emission levels or safety performance are poor or perceived to be so. Rightly or wrongly, regulatory officials are being obliged by events or encouraged to become more distant and independent in their relationships with managements. The many new tasks for HMIP which will result from implementation of IPC will undoubtedly restrict the time previously available for plant inspection and discussion of problems. The NRA correctly emphasizes its position as an environmental guardian which will rigorously enforce all criteria on liquid effluents and will not hesitate to take court proceedings where they feel they are justified.

In these circumstances, managements at the plant and at the board level can no longer rely on an understanding acceptance of their explanations for problems which inevitably arise. Conscious and continuing efforts are required to establish and maintain external relationships. Local liaison committees are usually helpful and a controlled 'open door' policy will usually repay the considerable effort entailed. Regulatory relationships may well become more difficult to maintain under the new regimes which are developing. Nevertheless, the joint interest in arriving at mutually acceptable public authorizations should promote the professional aspects of this relationship. Confrontational attitudes would best be avoided. Management attitudes and performance in these increasingly important areas form a fully justified part of an environmental audit.

The Implementation of Environmental Audits

The first stage of implementation is for management and auditors to agree the facts. It is also highly desirable to agree recommendations although these recommendations must ultimately be those of the auditor who is usually not in possession of all the facts – financial, marketing and strategic – surrounding a particular plant. Thus, the recommendations are developed correctly only from an environmental, health and safety standpoint.

The most difficult issues are prioritization and the related question of developing an action plan. Many relatively minor issues can and must be resolved at plant management level. Where significant capital expenditure is judged necessary, the problem is to decide how detailed and accurate the cost estimate must be and what priority it commands relative to the many competing demands facing most engineering departments.

There is no 'correct' answer to this problem. One approach is for the auditor to issue the audit, factually agreed, as quickly as possible and for senior management to allocate any cost estimating priorities which are necessary. Implementation must be a board matter since only at this level can the necessary capital proposals be allocated priority in comparison with other capital and strategic questions. If environmental auditing is given its proper status, the auditor should be allowed access to top management equal to that of plant management. This is to ensure a 'second chance' for the auditor to revisit any decisions which may have resulted from misinterpretation or lack of clarity in the report.

Follow-up of individual environmental audits must become part of routine management. An important aspect of this is routine contact between the auditor and plant management, as opposed to 'set piece' environmental audit visits. Nevertheless, in the rapidly changing world in which the process industries operate, it is advisable to perform repeat audits at, say, biannual intervals.

THE IMPLEMENTATION OF BS7750

During the progression of BS7750 from the published draft stage through piloting to final specification, the scope of the standard, certification, bureaucracy, the need to demonstrate continual improvement and the relationship with regulatory criteria have been discussed within the working groups and BSI committees.

The Scope of the Standard

Some flexibility regarding the scope of an audit exists provided that agreement can be reached with the certifier. The key questions are: (a) Is health and safety included? (b) Are major accident hazards included? (c) How far back (into raw materials) and forwards (into products) should the Environmental Effects Register be taken? and (d) What 'modular unit' within a company should be audited?

For BS7750, health and safety issues should probably be excluded though they are often included in internal company audits. Since the majority of audits are likely to be on a site basis, the Environmental Effects Registers required under the standard will probably refer only to those raw materials over which site management has control. Likewise, product issues will probably be restricted to packaging, warehousing and the first stage in the distribution chain.

Registers of Environmental Effects and Regulations

Great concern has been expressed, particularly from small to medium-sized companies, about the need to maintain a Register of Environmental Effects and a Regulations Register. As a rule, smaller companies do not possess the library facilities and specialist staff of large companies. There is no easy answer to this problem, though trade associations and numerous environmental databases will be valuable sources of information. Some EC legislation, such as the directive requiring data on existing substances, covers the information required for the Environmental Effects Register. On the legal and regulatory side, the complexity of the burden of legislation under which UK companies operate has proved surprising to many companies.

Certification

Although it encapsulates good management practice in a field from which, in modern circumstances, industry has no escape, an environmental management standard like BS7750 is considerably more complex and difficult to focus than a more straightforward 'quality' standard. Companies raise many questions about how certification is to be achieved. Some questions concern certification criteria. Is it a judgement of the efficiency of the management systems in place to deliver the stated environmental policy? If so, will the certifier be competent to judge compliance against quantitative criteria and, particularly, the gravity of any adverse effect caused by non-compliance? Other questions concern judgement of performance based on non-quantifiable factors such as externally perceived noise, malodours, raw materials purchasing policy and levels of recycling. Will it be sufficient for the management system merely to identify such issues rather than to make a necessarily subjective judgement on them?

THE FUTURE ENVIRONMENTAL MANAGEMENT STANDARDS

Complaints have inevitably arisen about the bureaucratic burden of BS7750, particularly for smaller companies. However, the management systems approach embodied in BS7750 will undoubtedly improve environmental performance, particularly if it identifies problem areas and provides at least the direction for operational solutions. Maximum benefit will be achieved if the audit part of the management system does not concern itself solely with compliance and does not become a top-down system imposed by external specialists.

Perhaps surprisingly, continuous improvement has not yet become a major issue. Some have suggested that continuous improvement is ultimately a recipe for zero emissions with potentially enormous costs. However, there are always likely to be areas for environmental improvement, but by no means always in terms of quantitative criteria. The continuous improvement principle, applied sensibly, should not be too onerous a requirement for industry.

It is not clear that convergence is developing internationally on environmental management systems, such as BS7750 and EMAS, in a manner which will minimize adverse environmental impacts. Nevertheless, such systems can be applied, to an appropriate degree, by large and small operations. No matter which regulatory system applies, the standards now being developed will provide a rational and scientific basis for the design of operations and management.

9. Infrastructure evolution and the global electronic marketplace: a European IT user's perspective

David Alexander

World markets, albeit depressed, are in a general state of flux. Compared with a decade or so ago, the market is now much more diverse, discerning, selective and demanding of value for money. Today's successful business enterprise has to be increasingly flexible and responsive. It has to reach the market more quickly with quality products that are differentiated from those of its competitors by particular added value specifically tailored to meet customer needs.

Businesses must also respond to challenges from the reshaping of political and trading frontiers around the world. In this context, there are a number of factors in play and history would suggest that each, in its own right, has potentially a major influence on world trade. In Europe, for example, the impact of integration initiatives in the European Community (EC), the unification of East and West Germany, the break-up of the Soviet Union, and, not least, national initiatives such as plans for massive privatization of French industry and commerce is making itself increasingly felt. The opportunities and threats ensuing from the greater openness of once protected national markets can be expected to be considerable in terms of both acquiring and defending market share.

To be in a position to respond to these dynamics and to play a part in revitalizing the world economy, businesses are being told repeatedly by leading economists that they need to pursue productivity improvements of not just the usual 10 or 20 per cent, but of a whole order of magnitude. The trend is now towards much more devolved decision making structures. The current emphasis is on down-sized firms that 'outsource' those operations that external specialists ought to be able to do better.

TECHNOLOGY FACTORS – EMERGENCE OF THE ELECTRONIC MARKETPLACE

With these immense pressures for change, businesses are beginning to look towards Information Technology (IT) not just as a tool for automation of existing processes but much more as an enabler and integral part of the required changes in business approach. Furthermore, as a consequence of the increased application of IT in this way, businesses are becoming more and more dependent on it for all operations. At the same time, the technology is creating new and highly efficient ways of doing things which in turn fuel increased market dynamics and even change the way the market operates.

Developments in Electronic Data Interchange (EDI) are a simple example of this. Where large customers insist on the use of EDI, suppliers have no alternative but to comply. Not only has EDI provided the transaction method, it has also triggered a new set of business precepts. 'Just-in-time' delivery, for instance, has major ramifications for all aspects of the production/supply loop.

EDI is merely indicative of what lies ahead. The new styles of doing business are demanding more comprehensive electronic trading practices and interworking methods. Increasingly, these will be needed for improved and more complex support of customer relationships, outsourcing partnerships with suppliers and increasing numbers of strategic and tactical collaborative ventures.

ELECTRONIC MARKET INFRASTRUCTURE – BY GOOD LUCK, DECREE OR EVOLUTION?

The electronic infrastructure will create a new geography of independent business opportunities. These will be based on the effective networking of a wide range of resources which will synergistically build on human and organizational relationships and will build and draw upon information sources around the world.

In its earliest phases, as well as providing for basic inter-enterprise business communication, the development of the infrastructure will need to be firmly targeted on eventual full and open interworking between IT systems, underpinned by seamless global telecommunication services. It will need to be based on international standards and backed up by methods which will ensure conformance of components. The system must work and continue to work as it grows.

The infrastructure will need to provide the common base for operation and interoperability, not just for large enterprises but also for medium and small-scale businesses and domestic consumers. The latter are becoming increasingly important components of an emerging networked economy involving home work stations, small office systems, mainframes and public and third-party services.

With this mix of systems and dependency on the interoperability of a potentially vast range of uncontrolled software components, it is inconceivable that the required infrastructure could simply be decreed, planned and built. A more realistic scenario would be natural evolution towards an infrastructure derived from the coalescence of interconnections of growing numbers of individual pieces.

If this thesis is correct, the essential electronic infrastructure could only emerge from the otherwise chaotic structure if a number of prerequisites are put in place. These include: (a) a comprehensive and sufficient range of products which are based upon international standards and tailored to be relevant to real business requirements; (b) widely accepted architectures to frame their effective deployment; and, most importantly, (c) an educated market with respect to the selection of the pieces.

INHIBITORS AND OPPORTUNITIES FOR PROGRESS

Some businesses are beginning to see the opportunity for a radical rethink of their business/IT/organizational relationships and, hence, there is a shift in emphasis towards infrastructural support for external, as opposed to internal relationships. Nevertheless, IT professionals are still tending to remain focused on yesterday's solutions and skills for yesterday's internal problems. This situation, perhaps caused and fuelled by recessionary pressure for reduced costs and rationalization, encourages incremental development of the already familiar proprietary systems and gives little or no lead to the market regarding the eventual standards-based solutions that will be required.

Compounding this, it is not purchasing in IT departments, but direct end-user purchase of IT solutions that is becoming the major shaper of the marketplace. Work stations now offer low-cost and powerful computing that falls within budgets for home use, for small enterprise IT applications and even for user groupings in larger enterprises. Often bought for specific immediate purposes and, as some suppliers claim, driven by fashion rather than function, such purchases tend to be undemanding of *de jure* standards. They provide little prospect of general interoperability and give no infrastructural lead to the market.

Putting to one side the problems of performance and compatibility which will need to be addressed in the next year or so, a fundamental issue stems from the current high cost of basic telecommunication which, if not checked, could severely distort and probably inhibit natural development of the electronic market infrastructure.

Even with today's limited network application requirements, cost is already an inhibiting factor. This is particularly true for small to medium-sized enterprises in countries, including the UK, where monopolistic control, or its residual effect, remains strong. Moreover, it is true for all enterprises operating internationally – a matter of particular concern for Europe where distances of a few kilometres can attract high international tariffs.

For the future, the situation could get very much worse if network services and tariffs are not designed to accommodate new types of business network applications. Networks will need to be able to efficiently cope with the short bursts of high bandwidth demand by these applications at costs which should remain low to reflect the expected low utilization of modern network resources when averaged over time.

Tariffs based on peak available capacity over time might be expected to generate massive revenues from emerging requirements. However, it is a simple fact of business life that costs will be constrained to stay below the 'ability to pay' threshold. Although this threshold may not be directly quantifiable, it is a state which is easy to recognize, as many businesses are finding, when their high telecommunication costs become comparable with their profit margins.

THE IT STANDARDS ISSUE – THE VICIOUS SUPPLY–DEMAND CIRCLE

Products based on *de jure* international standards must form the basis of any electronic market infrastructure. To date, however, the market for standards-based products has been slow to develop. The reasons are clear to see.

Business users of IT tend to be rather sceptical about standards development. The promise of open systems has not been delivered. Even if users believe that these standards might eventually be delivered, they remain cynical as to their likely value because of early products which were limited in scope, inefficient and difficult to use. The end result is that businesses tend not to build standards solutions into their system plans or procurement strategies, preferring to wait and see.

Furthermore, even when asked to provide input into the standardization process, business users are reluctant to get involved. They judge such activity, at best, as a waste of time and, at worst, as a potentially massive drain on scarce resources, should their key IT staff get involved in trying to influence even just a few of the hundreds of committees around the world.

On the other hand, during the last few years, many of the world's leading IT suppliers intensified their voluntary initiatives in the many standards committees. At the same time, however, they invested heavily in their own development of products based on *de jure* standards. Now they find that users have become reluctant to buy what they had seemed to be asking for in the 1980s. Instead, they now seem to be reasonably happy buying a mix of approaches to meet immediate needs. Clearly, the requirement has been moving on and the swifter, *ad hoc* initiatives producing *de facto* standards have been more able to keep up. Also, if they are at all aware of the transient and potentially tangential nature of such approaches, users now seem more prepared to accept the risk.

The fact that the market has not developed as fast as expected leaves the suppliers with a dichotomy. They could invest further in *de jure* standardization and related products or they could concentrate their main efforts on producing proprietary solutions. New leaders in the pack seem to favour the latter. Whatever the reason, it has to be concluded that the standardization processes for IT have not kept pace with advancing requirements and, perhaps not surprisingly, even seem to be unaware of the advancing changes in these requirements. From an industry perspective, this is a situation that cannot be allowed to continue if major risk is to be avoided.

In the current situation, there are few standards-based products, few buyers and, consequently, there is little product investment. The breaking of this classical and vicious demand–supply circle is not in the hands of any one party but needs clear vision and the cooperative partnership of all concerned – suppliers, standards organizations, and, most importantly, the users.

BUSINESSES AS USERS AND THEIR STANDARDIZATION ROLE

To understand better what the relationship of the business user to the standards process ought to be, it is useful first to view this process as a critical part of the demand supply loop shown in Figure 9.1. This way it can be seen that business users of IT are not the actual users of standards. It is the suppliers who use standards to design and build conforming products and services. Users will only buy and use their products and services if they satisfy business requirements.

In this context, it needs to be understood that user choices will necessarily be pragmatic. Business solutions that work irrespective of whether or not the solutions are standards-based must be found. It follows that standards-development must be keenly targeted to yield products and services that will meet real business needs and be available in appropriate time frames.

Only business users are in a position to provide the fundamental standards-related inputs on what their real requirements are likely to be and where their

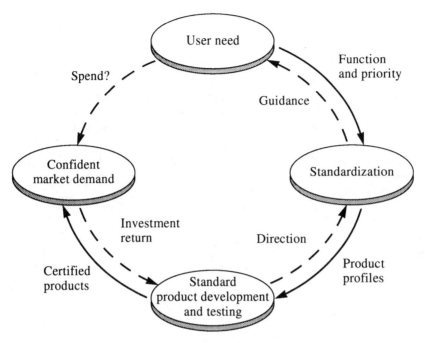

Figure 9.1   The vicious supply–demand circle

priorities lie. It is this information regarding their requirements that users should be asked to contribute to the process, established at a functional level and expressed in business application terms. Their scarce IT resources should not be dissipated by being drawn into the 'how' of standards – this must be left to the suppliers.

There are existing examples of how this could work. There are government-supported Open Systems Interconnection (OSI) initiatives for procurement like Government Open Systems Interconnection Profile (GOSIP) in the US and the UK, and the European Procurement Handbook for Open Systems (EPHOS) on a European scale. There are also the independent mission-driven activities of the European Group on Technical Office Protocols (OSITOP) and the related supplier-affiliated User Council of X/Open. However, because of the ingrained user scepticism of standards, only narrow segments of the market can be counted on to come forward in this way. The majority will need to be convinced by sound arguments that to do so is essential to the preservation of their business interests.

Consequently, it is not the technicians in an enterprise who will need to be convinced, but the high-level business managers. In this way, support and commitment will be assured in terms of the initial phrasing of requirements and also the setting of plans against which eventual purchases will be made.

Steps must be taken to ensure that the *de jure* standards processes and associated product development cycles are focused on, and responsive to the real and rapidly changing business requirements of end-users. Figure 9.2 suggests a general operational framework that positions fundamental user involvement, as part of a regenerative supply–demand loop at the levels of direction-setting and market-testing of the results. Stimulus and help may need to be given through industry employment of user spokespersons to act as catalysts and as translators of business requirements to functional specifications.

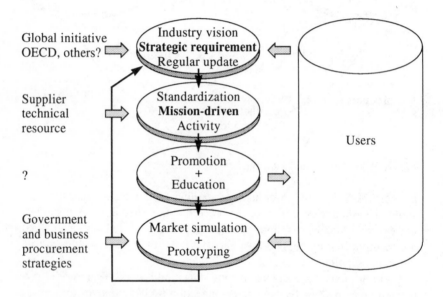

Figure 9.2   Market-focused standards framework

A global electronic infrastructure for the rest of the 1990s and beyond cannot be expected to result from chance. Telecommunication operators, third parties and governments should explore all options and opportunities for early and effective basic infrastructure support, particularly regarding lower tariff arrangements, both nationally and internationally. Users must play a critical market-shaping and standards partnership role as a prerequisite for progress in infrastructure development, and widespread support and commitment must be gained at the highest management levels of industry and commerce.

10. User involvement in the life cycles of information technology (IT) and telecommunication standards

Kenji Naemura

It has often been repeated that technical standards are important in order for information and communication systems to be employed most efficiently and effectively. Such systems cannot be implemented and interconnected as part of an infrastructure for the 'information age' unless suitable sets of standards are developed and utilized. It has never been completely clear, however, what types of standards are meaningful in this context and how they are important to the users of the systems.

LIFE CYCLE OF STANDARDS

Much has been said about the development of standards, but very little has been written about what becomes of the standards once they have been developed. While statistics are widely available concerning the standards which are under development and those which have been published, there are practically no statistics readily available concerning the systems that employ the standards.

Unlike conventional publications such as novels, textbooks and handbooks, printed standards do not have any artistic, intellectual or economic significance in themselves. Needless to say, standards exist to be implemented in actual devices, systems or services which are then offered to users in one way or another so that the users can benefit from their utilization. This is the only means by which standards can have a practical meaning as well as an economic value.

The above observation is true for all standards (including mandatory ones), and crucial for voluntary standards if they are to have any significance. Standards developers, therefore, should gather information on the actual application status of the standards; otherwise, they are selling products of unknown market value. Users should also request such information either from the system/service suppliers or the standards organizations; otherwise, they run the risk of purchasing non-standard products.

Most standards in the Information and Communication Technology (ICT) field have application periods of varying lengths. The average lifetime is shortening because of rapid changes in underlying technologies as well as the never ceasing emergence of new applications. Standards must be enhanced, updated or replaced accordingly; otherwise the systems using them would not be accepted by the users in the changed environment.

Thus, it would be more practical to define standards, not in terms of their publication status, but in terms of their life cycles: starting with creation and going through publication, implementation, modification, adaptation and so on, and terminating with cancellation or replacement. Moreover, each published standard should be supplemented with frequently updated information on its life cycle.

OFFICIAL STANDARDS VERSUS *DE FACTO* STANDARDS

Some standards are officially planned, developed and approved in recognized committees according to formal procedures. Others are created either in firms, groups of firms or in industrial associations. The former, often called official, committee or *de jure* standards, are naturally regarded as having a higher status than *de facto* standards.

This demarcation, however, may not be as clear as it was previously. As computing, telecommunication, broadcasting and so on converge, and we enter an era of multimedia and competition on a global scale, the value of all standards depends more and more on their market acceptance rather than the result of voting in a committee. The market changes very rapidly and often in an unexpected manner, and it does not usually care whether the applied standards are official or *de facto* as long as they exist and meet the needs.

On the other hand, it would be absurd to think that official standards were not required any more. There are various ICT areas in which a large number of technical items should be agreed upon officially before the actual implementation and deployment of systems begin. Otherwise, there would be a chaotic race between incompatible products. Some examples are network and terminal connectivity, interoperability, radio applications, software portability and information coding.

It is therefore necessary to consider the meaning and importance of both official and *de facto* standards, but with different perspectives. The criteria for defining the role of the different types of standards may not be simple, but they must include an assessment of the degree of 'network externality', the status of market competition and the degree of technical advancement of the standards in question. It should be noted that at some phase in its life cycle, a *de facto* standard may become an official standard.

STANDARDS VERSUS INTELLECTUAL PROPERTY RIGHTS (IPR)

More and more frequently, it is necessary to take account of IPR when implementing a standard. It becomes even more important in the case of *de facto* standards when the key IPR belong to a particular owner. Most of the standards organizations deal with IPR according to two principles: (a) before a standard can be approved, the holders of IPR essential to that standard are requested to submit a written declaration that the IPR will be licensed to users of the standard under reasonable terms and conditions; and (b) a standard will not be approved in the absence of such a declaration. The development of *de facto* standards, however, may not be based on similar principles and, thus, the users may not be able to apply such standards without becoming dependent upon the IPR holder.

Sooner or later, IPR concerns will make it necessary to define a new criterion to differentiate those standards which can be applied on a non-exclusive and non-discriminatory basis from others. The former may be called 'open' or 'public' standards, while the latter are 'closed' or 'proprietary' standards. Users should be fully aware which of these two classes of standards they employ. If they choose the latter class, they run the risk of being locked in by a particular vendor and having difficulty in switching to other vendors in future. On the other hand, proprietary standards should not be ignored altogether – as noted above, it is possible that a 'closed' standard may become an 'open' standard at a certain stage in its life cycle.

Standards and Users

There is no obvious answer to the question 'Who are the users of ICT standards?' Some users have more advanced technical knowledge or more comprehensive applications of the technologies than others. Some of them are professional engineers while others are bankers, retailers, personnel managers and so on. User applications differ and there are many varied user requirements, as well as some common ones. It may be necessary to separate personal users from business users in order to identify requirements of different kinds.

For any particular technical standard, it is possible to identify an expert or a group of experts who made the main contributions to its development. This expert or group can be called the provider of the standard. There will then be a number of product implementors who will deploy the standard as part of their product specifications. In the strictest sense, they are the real users of the standard, and, since they should be in the best position to assess its quality and deficiencies, their voices should be heard by the providers of the standard.

The above statements may seem rather straightforward and simple, but they are not. The dramatic progress in ICT has changed formerly centralized environments into distributed ones. The conventional notion of 'implementors as engineers' as employed by well-established ICT firms is not always valid any more. Many of the standards in these areas are actually implemented in small distributed computers, work stations and personal computers (Naemura, 1992b). The 'implementor' is becoming a much more diverse entity than ever before.

Therefore, it is necessary for the standards provider to assume that standards will be used by various implementors in different environments, and to try to identify these implementors as accurately as possible. It may be difficult to identify them, but they will decide the market value of the standard in the first instance.

It has been pointed out frequently that product implementors must be differentiated from the 'end-user' of the standard. In some cases, the implementor's product (and the standards used) become part of a chain. A product may be incorporated by a second implementor before being delivered to the end-user. Furthermore, an implementor of a standard may be in a position to provide this product to a second implementor employing another standard, who, in turn, will provide the product to a third implementor, and so on.

It is useful to consider the various types of users as positioned in a layered model analogous to the OSI Reference Model. The principle is that an entity in layer 'n' makes use of a service provided by an entity in layer 'n–1'. Here, a service may be a standard, a product or anything offered by the entity in layer 'n–1'. A standards provider would be located in the lower part of this model, and the users of standards in the higher part. The total number of layers need not be fixed, but in order to demonstrate the applicability of the model, let the number be seven as configured in Table 10.1.

This conceptual model is useful in clarifying general relationships among various players. For example, an end-user could ask a corporate information system manager for certain new functions. The manager would translate the user requirement into a system requirement, which would then be translated by a system integrator into a component requirement, and so on, until a suitable standard is developed and provided in the form of the real system as required by the end-user.

In some cases, an end-user is also responsible for corporate system management and system integration. In this case, the user covers the three upper layers and interacts directly with component product implementors who will

Table 10.1   Layered model of user types

Layer	
7	End-users
6	Corporate system managers
5	System integrators/network operators
4	Component/systems implementors
3	Functional standard providers
2	Base standard providers
1	Underlying technologies (e.g. semiconductors, software)

supply him or her with products. The layered model is also useful in coping with the increasing demand for private networks. These networks will be operated by organizations who represent the end-users to the implementors located in a lower layer of this model.

The model also exposes old and still outstanding problems. Who should pay for the service provided by the standards provider? Should the end-user be responsible for it? Should the standards developer be paid based on the copyright licensing or the patent licensing?

THE ROLE OF USERS

End-users have to realize that they stand to benefit from ICT standards and also that the costs of standardization are eventually included in the price of products and/or services. They must therefore regard themselves as the ulti-mate sponsors of standardization activities. Such a conceptualization will help users to be aware of their real needs, and to learn about existing stand-ards and requirements for new standards (Naemura, 1992a). The users' role, then, is to inform suppliers that the products and services they will buy should be based on specific relevant standards. Corporate systems managers will be able to formulate requirements priorities. This information will be very useful for standards developers in determining their time schedules. Users buying ICT products and services without requesting compliance with standards are doing so at their own risk.

Once use of the products or services has begun, users should evaluate the standards. They should also estimate how much economic value the stand-ards have realized in their environment. This estimation should take into account various factors that help reduce the user costs, for example, the number of possible choices among compatible products, the number of hardware and software products that could be connected to the new system and reduced amounts of required re-education.

After the evaluation is completed, the users will quite possibly find various points for improvement. They are now in a position to participate directly in the making or improving of some of the standards. In order to facilitate user participation, some improvement in the organization and methodologies of standards-setting may be needed. Separating external functionalities from internal specifications in the content of a standard may be necessary. Utilization of electronic mail and other electronic means for discussion and the exchange of proposals should also be considered.

Users should be encouraged to contribute to the costs of developing standards. Large users can and should be members of standards organizations and pay the necessary member fees. Smaller firms may incur smaller costs by participating through industrial associations. Academic and individual users may contribute by offering their own time and certain skills to carry out editing and other associated works. In many cases, users may indirectly participate in standardization activities by delegating the activity to other large users, specific suppliers, consultants and so on. To be effective, this delegation should be clearly identified.

ORGANIZATION AND PROCESSES FOR USER INVOLVEMENT

Basically, standards development should be market-driven. Many of the modern standards organizations have mechanisms to search for user requirements. In some technical fields, countries and regions, these mechanisms need to be strengthened so that user requirements and user evaluations can be put into the process adequately.

The Third Inter-regional Telecommunications Standards Conference (ITSC-3) held in Tokyo in November 1992 was perhaps one of the first occasions for the leading standardization organizations to discuss extensively the issue of user involvement. Contributions were received from the three founding members of the Conference: the American National Standards Institute 'T1' Committee, the Japanese Telecommunications Technology Committee (TTC) and the European Telecommunications Standards Institute (ETSI). Additional contributions were submitted by the Telecommunications Standards Advisory Council of Canada and the Australian CCITT committee.

In addition to network operators, the membership of each of these organizations includes users, manufacturers and other types of members. In theory, it should be possible to develop standards satisfying everyone's requirements. The reality, however, is that there are two basic impediments:

1. that user members of the organizations do not always actively contribute to the process; and
2. that the vast majority of users are not members.

Clearly, some means should be found for the organizations to facilitate user involvement and to ensure that the standardization process will take their views into consideration.

The T1 view in this regard could be summarized by the following four principles (T1, 1992).

1. Openness: standardization activities and plans must be made as widely known as possible.
2. A multi-staged approach: this consists of a project approval stage at which user input is specifically sought, and a three-stage process that becomes more technically specific as standards development moves closer to technical implementation.
3. Availability of standards: the information contained in the standards must be made available to users through a variety of media.
4. Diverse opportunities for user involvement: this is achieved through direct membership, indirect membership (input without obligations), and through interactions with other organizations related to the standardization process.

The TTC approach is similar to that of T1 but with a strong emphasis on special subcommittees and procedures for collecting user requirements (TTC, 1992). A Subcommittee for Hearing User's Requests investigates possible user requirements at each stage and inputs them into the standardization process. It also conducts surveys to find other requirements from non-members. The results of the surveys are submitted to TTC which deliberates on standardization planning and coordinates overall activities. The TTC Five Year Standardization Project Plan is updated and published as a result of such deliberations.

There is also a special procedure for obtaining opinions and requests of TTC members even before the establishment of a standardization project is proposed. This is meaningful when, for example, certain services are planned by a network operator which are not necessarily subject to standardization, but are somehow related to it. Although this procedure is not aimed at users, it does encourage them to express their requests as well as to educate standards developers to listen to the users.

In fact, a similar approach has been taken in a more comprehensive manner by Japanese standardization organizations in areas other than telecommunication. In 1992, for example, the Japanese Industrial Standards Committee

(JISC) (consisting of members representing suppliers and consumers as well as neutral experts) established a special subcommittee to discuss future relations between standards and the personal lives of individuals. All sectors of society were represented. Several issues came up that were related to ICT systems – compatibility of software and printer supplies, procedures for using the equipment and terminology, electromagnetic interference, and so on – and JISC has channelled these user concerns to appropriate technical subcommittees.

ETSI is also concerned very much with ways to facilitate user involvement (ETSI, 1992). In 1992, its Secretariat committed significant resources to addressing this concern and conducted over 50 interviews with user groups and other user enterprises both inside the ETSI membership and outside.

The ETSI survey found that users do not necessarily understand why they should have to pay for memberships and participate in the standardization process. In terms of money, time and human resources, the cost of directly participating in the process is too high for them to justify. However, users would like to play a part in setting standardization priorities and to be informed ahead of time when their input is needed. Naturally, they are interested in what the standardized product or service is going to do for them and not how the product or service achieves this result.

The ETSI Secretariat submitted some proposals to its Technical and General Assemblies aimed at affecting fundamental changes in the way user participation problems are addressed. The first element of the change would be behaviour: that all people involved in standardization work recognize the difference between technical experts and user experts, that they pay special attention to the way they communicate, and that ETSI must be open to actively seeking the views of non-member users. The second element would be tools: the ability to flag where user input is needed and to communicate with the widest user population. The last element would be to establish a new structure called a 'User College' in which ETSI members (and others) discuss issues and coordinate actions to ensure that the user voice can contribute to the success of standards.

The Australians contributed a comprehensive study of the user issue to ITSC-3 (Australian CCITT Committee, 1992). It identified several reasons why user input has been traditionally overlooked – the engineering-driven nature of the domain, the lack of necessary knowledge outside the telecommunication industries, the lack of regulatory mechanisms by which end-users could make an input into standards-setting, the lack of public awareness about the importance of standards, greater market differentiation causing diversified requirements, and the lack of identification with standards-setting bodies.

The Australian study suggested a number of key requirements to structure user input:

1. policy (preferably through legislation) to protect the interests of users in the standards-setting body;
2. structures to allow public consultation and input;
3. establishment of a standards register;
4. establishment of electronic bulletin boards;
5. use of consumer-friendly language;
6. use of a *pro forma* response sheet for regular surveys;
7. use of existing networks and outlets to publicise activities and solicit input;
8. getting user organizations to set priorities; and
9. constant review and assessment of initiatives for obtaining user involvement.

To summarize the ITSC contributions, there was a fair amount of consensus on the nature of the problem and the policies and mechanisms necessary for solving it. Points of common concern included:

1. the importance of informing the users, in words they can easily understand, of the role of standardization and the ways in which they can get involved;
2. publication of documents in a manner that encourages users to respond with comments and requests; and
3. prioritization of standardization projects such that they seek user inputs and clarification of user objectives.

Although some differences remain regarding the emphasis placed on certain elements of the mechanism, it is expected that a harmonized view on this matter will be achieved in the newly created Global Standards Collaboration Group (GSC), the successor body of ITSC. Its result will also benefit the discussions in the Telecommunication Standardization Advisory Group of the ITU Telecommunication Standardization Bureau.

Immediately following ITSC-3, a separate Workshop was convened, jointly organized by the Japanese Ministry of International Trade and Industry (MITI), the Ministry of Posts and Telecommunications (MPT) and the Organization for Economic Cooperation and Development (OECD). Its topic was 'The economic dimension of IT standards: users and government in the standardization process' and the participants represented various interests from several countries. There was some difference of opinion, but there also seemed to exist a general consensus on the current status and problems regarding user involvement in standards.

In addition, there were various organizations either represented or mentioned at the Workshop, which did not belong to the traditional circles of standards-setting bodies but which contribute to the promotion, application, stimulation or evaluation of standards. To some extent, these groups play an intermediary role between standards-setting bodies and users. They include business groups working on the application of Electronic Data Interchange (EDI) standards, regional Open Systems Interconnection (OSI) workshops, as well as fora on network management, Integrated Services Digital Network (ISDN), Asynchronous Transfer Mode (ATM), and others.

The Workshop came to several important conclusions. Firstly, it was generally agreed that ICT standardization has been technology-driven, but that it should be more market-driven. To achieve this goal many different types of users have to find appropriate methods of representation. Too often user views are 'filtered' by a hierarchy of supplier-dominated committees. Secondly, it was agreed that the timing of standards was very important. The process often takes 5–10 years, and should be shortened to about 3 years. More of the strategic and economic considerations in standardization should be reviewed at management levels in the member organizations. Finally, it was acknowledged that there is a need for better research into how standards organizations work and that there is a lack of university-level education on the importance and the workings of the standardization process.

ISSUES FOR FUTURE STUDY

Some of the above issues will need relatively long-term study, involving joint efforts among researchers with backgrounds in economics, political studies and engineering. Methods need to be developed for the evaluation of standards from the user viewpoint. A means must be established to define clearly the user requirements in standards development, including government procurement requirements. A new methodology needs to be developed to explore the layered model of users and their requirements as well as to correlate them with the technical specifications.

Further analysis is needed of the effects on users of the protection of the IPR in standards. The effect of copyright as well as patents should be taken into consideration. The result of this work will possibly set a new demarcation between public and proprietary standards. In general, the qualitative and quantitative benefits brought to users by ICT standards must be clarified. Studies should take into account the kinds of standards and their life cycles, and the results should be expressed in a manner intelligible and persuasive to the users.

PART 2

Institutional practices and objectives

11. Standards and standards-setting processes in the field of the environment

François Leveque

INTRODUCTION

This Chapter reviews environmental standards and standards-setting processes using an analytical framework which emphasizes a firm's incentives to adopt an environmental standard and the informational and institutional factors which influence the efficiency of standards-setting. It is argued that standards-setting processes, like any other regulatory process, strongly affect the effectiveness of the policy instruments which are adopted.

The Chapter begins with a presentation of the analytical framework. Firstly, standards are categorized according to the main economic function which they perform and the nature and the extent of the private incentives which they provide. Secondly, standards-setting processes are described as coordination mechanisms which reduce informational asymmetries and uncertainties, and embody institutional factors. Three forms of standards-setting processes are stylized: self-regulation, 'flexible governance' and command-and-control.

A discussion of the different standards-setting processes which lead to the adoption of quality, compatibility and social standards, follows. The main concern is the selection of self-regulation, flexible governance or command-and-control mechanisms for setting standards. The discussion is based on the previous analysis and empirical evidence, in particular that relating to green products and packaging of waste. The choice between standards and alternative instruments, such as economic incentives, is briefly considered.

PRIVATE INCENTIVES FOR THE ADOPTION OF ENVIRONMENTAL STANDARDS

The nature and the extent of incentives for firms to commit themselves to standardization in relation to environmental performance depend on the functions the standards perform. Literature on the economics of standardization (David, 1987) has identified two main properties of standards, namely the reduction of transaction costs and the exploitation of network economies (Kindleberger, 1983). These two functions provide *ex ante* private incentives: before the standards-setting process begins, firms may expect a gain provided by the market. Conversely, when standards deal with the reduction of pollution externalities, firms may expect to incur a cost, for example, for pollution abatement. This is the problem addressed by environmental and resource economics (Bohm and Russell, 1985).

Quality Standards

These reduce transaction costs (Foray, 1993b). The information embodied in such standards reduces informational asymmetries and uncertainties between the seller and the buyer with respect to the performance of the good. Incentives are provided through reputation mechanisms, with their level depending on the consumer's willingness to pay for quality. Generally, in relation to environmental quality, product performance is unverifiable: consumers cannot assess the environmental friendliness of products and processes. Trust therefore plays a major role in establishing incentives for a 'green' reputation (Brusco and Cottica, 1993).

Compatibility Standards

These assist in the realization of network economies (Katz and Shapiro, 1985). When a consumer's utility increases with the number of other agents consuming the same good, the adoption of such standards lowers costs and increases the variety of complementary goods, for example software, on the market. Like economies of scale based on experience, network economies are a function of cumulative output. However, they are less firm-specific than economies of experience in production (Gabel, 1987). Consequently, cooperation matters greatly. Incentives are provided through networking and their level depends on the exclusion of free-riders. In the field of the environment, compatibility standards relate mainly to waste management.

Social Standards

Those designed to protect individuals, for example those related to health, form another category (Kindleberger, 1983) which serve to reduce negative social externalities. Since the externalities are negative, firms are likely to oppose social standards. They may use their obstructive power to oppose regulation in order to avoid standardization costs. Moreover, the social dimension of these externalities suggests that the issue of legitimacy is at least as important as that of efficiency (Majone, 1978). The legitimacy of setting a new social norm is provided by moral and scientific arguments. Therefore, adoption depends on the balance between the legitimacy of the standard and the pollution abatement cost. Social standards are not supported by private incentives and their adoption is not rewarded by market mechanisms. As a consequence, self-regulation cannot be expected to emerge. However, public and government pressure which pose threats to industrial activities may then provide incentives to firms to adopt social standards.

Table 11.1 summarizes the nature and the extent of the incentives for the three categories of standards. Since standards are not the only available means, the main alternative instruments have been indicated. Two main objections can be levelled against this taxonomy. As pointed out by David (1987), in the real world a given standard can perform several functions. This may be true, but there is empirical evidence that standards generally perform one *main* function. Also, from an economics point of view, the function of a standard corresponds to a specific mechanism for internalizing externalities.

The second objection is that from the standpoint of orthodox economics, all standards, in the absence of government failure, produce an improvement

Table 11.1   Incentives associated with different types of environmental standards

Environmental standards	Quality standards	Compatibility standards	Social standards
Function	Reducing transaction cost	Exploiting network externalities	Reducing social negative externalities
Ex ante **private incentives**	Yes	Yes	No
Incentives mechanisms	Reputation effects	Networking	Credible threat
Determinants of the extent of the incentives	Willingness to pay for quality, trust in prescribers	Network economies, free-riding limitations	Pollution abatement cost, legitimacy
Alternative option	Moral suasion	Technological gateway	Economic incentives (taxes, subsidies, tradeable permits)

in social welfare. Quality standards limit negative externalities derived from asymmetric information relating to product performance, and compatibility standards lower negative externalities caused by the absence of technological interrelatedness. The taxonomy in Table 11.1 would be irrelevant from this point of view because it only covers, from an empirical perspective, different categories of market failure.

However, the approach adopted in this Chapter is different. Firstly, we assert that environmental standards are supplied in response to a social demand instead of being set to eliminate the unfortunate allocative consequences of market failure. Secondly, it is argued that social standards are set, not to correct a malfunctioning of the market, but to *constrain* the market; that is, to prevent goods from being regulated by purely economic forces. Thirdly, we assume that economic agents have a loss aversion, that is they are more sensitive to a loss than to a gain of the same magnitude (Kahneman *et al.*, 1991). Positive and negative externalities are not equivalent. In particular, bargaining is easier when externalities are positive (Willinger, 1993).

STANDARDS-SETTING AS A REGULATORY PROCESS

Stylized Regulatory Processes

Three processes for setting environmental regulations, defined in the broadest sense to cover all available policy instruments, are defined. These comprise two unilateral processes (command-and-control by government and self-regulation by industry) and a bilateral process (flexible governance between industry and government). These processes are stylized as follows (Leveque, 1993b).

1. The command-and-control process: this is a unilateral coordination mechanism for designing regulations in the sense that it is essentially an administrative process based on the coercive power of the government. Interest groups can be consulted, but they are exogenous participants in the process. A firm's rationale is assumed to be driven by regulatory capture (rent-seeking behaviour and obstructive power). Besides coercive power, government requires an *ex ante* knowledge of the pattern of the problem and the solution. The main constraints on the efficiency of such a process are rooted in administrative costs and failures, regulatory capture and innovational lock-ins.
2. Self-regulation (the unilateral commitment of firms): this is not a policy-making process in which the government is endogenously involved. Why should government regulate an industry if the industry itself is

'efficiently' and 'spontaneously' tackling an environmental problem? However, just as firms influence the command-and-control process, public authorities influence the self-regulation process. Either they play an invisible role, by wielding the threat of legal action, or they are consulted by industry and may even give approval. The prerequisites for self-regulation are: (a) public or administrative pressure which poses a threat to industrial activities; and (b) a positive market gain if self-regulation is established. The main obstacles to such a process are free-rider effects and the commitment of firms. For example, if the regulation is set up by a collective industrial organization, the coercive power of the principal (the industrial association) in relation to the agents (individual firms) may be very weak.

3. The flexible governance regulatory process: this involves a bilateral coordination mechanism between government and a coalition of firms. Industry, as well as other interested parties, is directly involved in the process. It helps to define objectives, the choice of means and the time-schedule for compliance. Industry therefore contributes significantly to the definition of the outcome of the process. The logic of this interactive process relates to the minimization of administrative costs, the identification of option values and the attainment of consensus (Van Vliet, 1992). Firms seek to reduce uncertainties and to gain stability and trust. The prerequisites are a credible threat on the part of the government and a positive sum game, even if the gains are intangible, taking the form of learning, economies of scale or scope, or an escape from more stringent command-and-control regulations. The main constraints relate to free-rider effects and lack of enforcement.

The rationale for this stylization is twofold. Firstly, the proposed taxonomy escapes the all-or-nothing view of public/private intervention in the setting of standards. As pointed out by Salter (1988), the policy issue is not simply whether or not government should intervene in the process of standards-setting. The real question is the combination of public and private involvement. The increasing number of voluntary agreements struck between industry and government (Glachant, 1993), such as in the case of Chloro-Fluorocarbons (CFCs), confirms this. Secondly, standards-setting is not different from establishing other regulations. Here, standards-setting is viewed as a regulatory process, that is, the phase which precedes the adoption of a regulation. The outcome of the standards-setting process determines the form which regulations take (Spulber, 1989). In relation to the environment, the choice of instrument is not always decided before consultation and bargaining takes place. The selection of the instrument may be one of the issues at stake.

It seems more appropriate than ever to interpret environmental regulations as the outcome of an economic and political process as opposed to economic stimuli to which firms simply respond (Skea, 1993). Path dependency due to increasing returns (Arthur, 1988), the dynamic feature of innovation (Franke, 1993), questions of legitimacy when uncertainties dominate (Majone, 1978) and eviction mechanisms in formal standardization (Foray, 1993b) are some of the theoretical arguments which stress the value of viewing standards-setting as a process. However, regulatory economics provides some more general reasons why processes matter in standardization.

Efficiency Issues

In a perfect world, as assumed within public interest theory, an *ex ante* assessment of the efficiency of a regulatory instrument can be made. However, in an imperfect world, where strategic behaviour prevails and information and transactions entail costs, procedures affect the efficiency of the instrument which they produce.

Interest group theory (Becker, 1983; Peltzman, 1976; Stigler, 1971) demonstrates that small homogeneous groups (for example oligopolistic firms) capture public regulations and are overrepresented because the stakes are higher and organizational costs are lower than is the case for larger groups (for example small and medium enterprises). Procedures to balance different private interests, including those of powerless stakeholders, are therefore required to attain greater regulatory efficiency.

New regulatory economics (Laffont and Tirole, 1993) based on the agency theory shows that the efficiency of a regulation depends on the incentive mechanisms set by government to collect private information. Because of informational asymmetries and uncertainties, the major inefficiencies relate not to the instrument, but to the incentive system integrated into the contract between the regulator and the firm and the way in which it tackles moral hazard and adverse selection.

A regulatory process is defined here as a dynamic coordination mechanism between economic agents which results in a set of collective rules to be enforced by each individual agent in order to tackle environment issues. It reduces *ex ante* informational asymmetries and uncertainties and embodies institutional factors (opportunistic behaviour and trust between the agents, and the coercive power of the government) which have a strong impact on the *ex post* efficiency of the instrument.

In terms of information, there is a general pattern in relation to the environment. Firstly, most of the information concerning pollution levels, abatement costs and technology is known only to the polluting firm. Secondly, the individual firm is not able to assess pollution impacts, especially in relation to the external costs of global problems. Thirdly, firms do not know, *ex ante*, the preferences of the regulator regarding the means, the time-schedule and enforcement rules. Fourthly, since several environmental concerns are new, scientific, technological, policy and organizational uncertainties dominate. As fora for exchanging information, environmental regulatory processes provide informational gains. If, for procedural reasons, these gains are limited, the outcome of the process is likely to be inefficient. There are numerous examples of bad standards leading to high compliance costs, costly enforcement, industrial flight or obstructive behaviour by firms.

Regulatory processes embody institutional factors to the extent that they enable each participant to assess the standing of others. This generates new rules. For instance, opportunistic behaviour may be observed, and then rules for limiting free-rider effects, such as the separation between ownership and the use of standards (Gabel, 1987), are introduced. Similarly, the confidence of the regulator in firms' commitment can lead to the modification of enforcement rules and thus, *ceteris paribus*, increase the *ex post* efficiency of the standard.

Effectiveness

Four variables influence the effectiveness of the different forms of the regulatory process:

1. *ex ante* private incentives;
2. *ex ante* informational asymmetries and uncertainties;
3. firms' opportunistic behaviour; and
4. government coercion.

To be effective, command-and-control requires coercion and only minor informational asymmetries and uncertainties. When coercive power is weak, regulations are ineffective since there is insufficient government power to enforce compliance. Obviously, if there is no credible government pressure, but there are private incentives, command-and-control is inefficient relative to the two other forms of process. When informational asymmetries and uncertainties are high, command-and-control is inefficient. The informational gain provided by this process is low and costly. Moreover, *ex ante* informational asymmetries and uncertainties permit opportunistic behaviour, and the obstructive power of firms can easily undermine the regulation when coercive power is weak.

Effective self-regulation requires high *ex ante* private incentives and limited opportunistic behaviour (or small informational asymmetries and uncertainties). When there are no incentives, firms do not regulate themselves. If incentives are combined with opportunistic behaviour, self-regulation is also likely to be ineffective. However, if informational asymmetries and uncertainties are low and opportunistic behaviour is constrained, self-regulation may be effective. Similarly, when incentives, asymmetries and uncertainties are high, but opportunistic behaviour is limited (that is, there is a high level of trust between firms), self-regulation may be effective since cooperation will prevail.

To be effective, flexible governance requires *ex ante* private incentives and a degree of coercive power depending on the informational context and the level of incentives. Flexible governance involves the government and an industrial association or a cooperative organization. In the absence of incentives, there is no reason for firms to cooperate with government. Coercive power is also required because, otherwise, self-regulation would be more efficient. The level of trust and coercion required depends on the informational context and the level of incentives. When incentives are low, the need for coercive power is higher to the extent that the expected benefits relating to cooperation are lower and, therefore, the pressure from the government to make firms cooperate must be higher. However, when both incentives and asymmetries are high, a greater degree of coercive power (or trust between firms) is required because, otherwise, cooperation would fail.

THE SELECTION OF A REGULATORY PROCESS

Environmental Quality Standards

Incentives and information
Incentives to adopt environmental quality standards are related to the willingness to pay for quality and the level of trust felt by consumers. There is no uniform willingness to pay for environmental quality. The environmental awareness of consumers varies according to educational and cultural factors. For instance, Scandinavians are said to be more aware of environmental concerns than the Spaniards, the French or the Italians. Nevertheless, polls carried out in Germany have shown that only a few citizens are prepared to pay higher prices for 'green' products.

Environmental performance associated with the energy and material content of some products can be verified by buyers before or after purchasing. Consumers may buy concentrated washing machine powder requiring less packaging material, window frames in wood instead of plastic or aluminium, wine in returnable glass bottles, long-lasting products and energy-saving household appliances. However, apart from these cases, environmental performance is generally unverifiable. Consumers cannot assess the environmental friendliness of products and processes. Informational asymmetries and uncertainties are low only to the extent that quality performance concerns a small number of conspicuous criteria.

Trust therefore plays a major role in incentives relating to a 'green' reputation (Brusco and Cottica, 1993). Product labelling set by firms or government renders environmental quality observable by consumers even if environmental performance remains *unverifiable*. To exercise their preferences for environmental friendliness, consumers are obliged to follow the advice of green groups, firms or public authorities. Institutional aspects are therefore very important because the credibility of these prescribers differs greatly from one country to another. In particular, environmental standards set unilaterally by firms may not be trusted by consumers.

Moreover, even for eco-labelling professionals, the environmental quality of a product is difficult or impossible to assess and verify, especially when several criteria are taken into account. There is no perfect objective index of environmental quality (Nadai, 1993). This reinforces the role of trust as a key institutional variable and implies that green standards can easily be challenged by competitors.

Quality standards-setting processes

The setting of environmental quality standards through command-and-control mechanisms cannot be expected unless there is only a single criterion at stake and incentives are low or negative. This is corroborated by empirical evidence. In the US and in the European Community (EC), the labelling of genetically modified organisms, such as tomatoes or potatoes, has been discussed. Biotechnology firms strongly opposed this quality standard. It would have strongly discouraged consumers from buying such products since public fears concerning genetic manipulation are very high. Similarly, there have been attempts to label strawberries and onions irradiated to lengthen shelf-life.

Mandatory quality standards tend to label environmental unfriendliness in cases where it is simple for an outside agency to exercise control and to collect information, namely when informational asymmetries are small. Therefore, command-and-control is not very costly and there is no private incentive for firms to accept such a standard or to set it up voluntarily, unless competitors produce a good which respects the standard and may consequently gain market share. Mandatory quality standards are very close to social standards-setting. The goal is not primarily to reduce transaction costs, but rather to reduce perceived negative externalities.

The rule requiring high incentives and low levels of opportunistic behaviour drastically limits the extent of self-regulation in relation to environmental quality standards. The rule implies trust and a high willingness to pay.

Empirical evidence shows that green product standards set unilaterally by firms relate to very sensitive issues and embody a single criterion. Dolphin-safe tuna, phosphate-free washing powder, chlorine-free paper and non-exotic timber furniture are classic examples. They reflect strong market incentives because of widespread public concern. Since these labels address only one criterion they are relatively easy to assess and manage without government intervention. The standardization costs are reduced and the non-compliance of a competitor is easy to detect and enforce.

Compared with self-regulation, flexible governance has three advantages. Firstly, it provides informational gains. Eco-labelling entails a collective discussion regarding environmental performance criteria. Criteria are based on the information provided by the producers and impact assessments carried out by committees in which industry participates. Such a process permits a large number of criteria to be taken into account. For instance, the French eco-label on paints is based on 20 criteria, and firms exchanged the results of individual life cycle analyses. Secondly, flexible governance enhances reputation since the label is guaranteed by a public institution, such as a standardization body. Consumer trust is thereby enhanced, to the extent that the public institution itself is reputable. Thirdly, the flexible governance process itself defends against loss of reputation. The assessment of environmental performance is based on consensus and compromise between the interested parties. When agreement is reached on the features of an eco-label, subsequent controversies – and thus the destruction of reputation by a competitor – are more unlikely.

In the recent past, several firms which have set up individual environmental quality standards have underestimated competition and trust issues. Competitors have denied the environmental improvement due to proposed standards, and controversies have led to a decrease in consumer confidence in standards supported only by firms. It is now common for firms to seek external support from government, scientists, green groups or consumer associations when standards are set. Eco-labelling, rather than self-regulation, is now the dominant mode of quality standards-setting within the EC.

Quality standards versus moral suasion

Moral suasion can take several forms: better environment awards for industrial schemes, lists of products and firms established by environmental groups and industrial charters such as the Responsible Care initiative of the International Council of Chemical Associations. Instead of adopting a product standard, firms adopt a commitment to do their best regarding environmental performance. There are three main differences regarding quality standards. Firstly, the agency problem of moral hazard between the seller and the buyer (that is, the buyer cannot assess the environmental performance of the firm) is overcome, not through a contract, but through an attempt to change the preferences of consumers by making them identify with the firm from which they purchase. Secondly, third parties such as green groups can act as watchdogs, controlling the voluntary commitment of firms. Thirdly, moral suasion provides much more flexibility.

The relative efficiencies of quality standards and moral suasion are very difficult to assess *in abstracto*. Moral suasion has analogies with self-regulation and is very dependent on the institutional context, especially regarding enforcement issues. Generally speaking, if monitoring a quality standard is technically impossible or prohibitively expensive, moral suasion may be more efficient.

Environmental Compatibility Standards

Compatibility standards are less relevant for environmental technology than for information and communication technology. Compatibility concerns in the field of the environment relate mainly to technological interrelatedness between waste products and services, that is, between waste and post-consuming activities such as collection, sorting, reuse, chemical and physical treatment, recycling, incineration and landfill. Zero compatibility occurs when a specific system is implemented for each product, for example the recovery and reuse of returnable glass bottles. Landfill and incineration provide almost total compatibility because they can use nearly all kinds of waste.

Incentives and information

Waste transportation is very expensive and waste resources are very heterogeneous. Network economies in collecting and sorting are high (Angel, 1994). As a consequence, a single waste-producing firm has no interest in implementing its own specific post-consumption system. On the contrary, it may be worthwhile, depending on the cost, for a firm to adopt a waste standard which is compatible with an existing recovery system. The compatibility standard can be close, or very distant, from the current technology the firm uses. Due to cost penalties, each firm would wish others to move closer to its own technical standard (Berg, 1989).

Compatibility standards tend to be undersupplied by the market (Katz and Shapiro, 1985; Farrell and Saloner, 1986). The incentives for a dominant firm to open up its private standard can be lowered by a loss of technological competitive advantage or market power. There may be greater profits from being a monopolist in a small market than an oligopolist in a larger market. Cost asymmetries regarding the integration of compatibility, equitable sharing of the burden and free-rider effects severely reduce the likelihood of cooperation (Gabel, 1987).

Undersupply of standards is less likely in waste management. In this sector, dominant firms possess the facilities to collect, sort and even recycle waste. Indeed, due to transportation costs and the high number of waste producers, these facilities have the characteristics of a natural monopsony. Dominant firms can easily constrain their suppliers to adopt compatibility standards which do not pose a threat to their market power. Besides, firm rivalry relates less to waste than to the product which generates it.

Governments have poor information about the implementation of waste standards, waste production is dispersed and heterogeneous. Waste producers have less information regarding the outlets of secondary materials than material producers. Finally, informational asymmetries are high between firms belonging to different material *filières*.

Many post-consumption activities, such as plastic recycling and sorting technology, are just emerging. The implementation of new waste facilities and recovery systems requires high investments which are difficult and costly to replace. Premature lock-in effects could be very inefficient. In this respect, cooperation provides information regarding the implications of standards and can decrease: (a) costs incurred due to Research and Development (R&D) rivalry; and (b) the risk that a firm will 'back the wrong horse'.

Standards-setting processes

Command-and-control mechanisms are required when private incentives are small or non-existent, and the standard has necessarily the feature of a public good. Safety requirements for industrial plants are a classic example (Hemenway, 1975).

As regards waste, two different contexts are worth examining depending on whether or not social standards regulate post-consumption activities. Until recently, post-consumption activities have developed mainly in a market context. The recycling of glass, paper, metal or wood was dependent mainly on the price of raw materials and the waste market was dominated by materials producers. The options for waste producers were landfill or supplying secondary material. If the latter was more profitable, they were obliged to adopt the specifications, for example, purity or minimum supply contracts, required by primary producers. External economies were mainly captured by a single firm or a small group of firms. Natural monopoly barriers due to waste transportation costs and the oligopsonistic market structure of raw materials rendered it unnecessary to restrict the standard. Government intervention was not required for establishing compatibility, but might have been for extracting monopoly rents. In the absence of social standards, government intervention was not required since, indeed, the market provided a high level of compatibility: landfill was the dominant technology.

Under new waste policies, the total compatibility provided by landfill is constrained. Several European countries recently have banned the landfilling of packaging waste, and have set social standards regarding the rate of recycling and incineration which should be achieved. There are three possible means to achieve this objective: (a) letting firms organize themselves; (b) setting a voluntary agreement between government and industry; or (c) designing a unilateral administrative scheme of planning and control. The latter option is clearly inefficient since the information required is private and very dispersed. Voluntary agreements are becoming the dominant pattern in most European countries, leading to the creation of monopolistic packaging waste consortia such as Eco-Emballage in France or Dual System Deutschland (DSD) in Germany. This alternative dominates because the participation of government induces a larger cooperative network, and therefore greater network economies, than the market would provide.

The effects of the first option – self-regulation – would be costly overstandardization, namely an undersupply of compatibility. Each homogeneous group of packaging firms would establish its individual waste system and compete for the limited capacity of consumers to sort waste. The packaging waste recovery and recycling industry would therefore be segmented into numerous *filières* and sub-*filières* specialized in different material/product combinations such as recovery and recycling systems of glass bottles, ferrous soft drink cans, aluminium cans and composite beverages packaging. At least a dozen waste containers per household, apartment block or supermarket would be installed, and consumers would spend much time sorting waste. The number of containers would equal the number of specialized systems which, in turn, would equal the high number of cooperative networks which self-regulation would provide. This is the current pattern in Italy, where numerous recycling consortia work independently. Self-regulation dominates because the government lacks credible coercive power to implement the European packaging waste directive.

An alternative to this scenario is the creation of a joint organization by all the packaging-producing and -using firms. This independent, but jointly owned company would be chartered, like Eco-Emballage or DSD, to set up an industry-wide monopoly system of recovery and recycling. The emergence of such an organization via self-regulation is very unlikely. The allocation of equity and the setting of differentiated prices according to the materials and products would be impossible to achieve. Informational asymmetries between firms and uncertainties relating to future prices of materials would impede such a collective agreement. Moreover, some producers have already implemented networks for collecting discarded packaging (for example, glass and paper packaging). There would be no incentives for them to join the new organization. Competitors specializing in other materials could not force them to participate.

The free-rider effect has been limited within the Eco-Emballage and DSD schemes. The ownership and use of the compatibility standard are separated. The standard is owned by the waste consortium while the green dot makes it easy to identify products belonging to firms which participate in the network. This provides a means to exclude the recovery of discarded packaging generated by outsiders. However, the financial burden is not equally shared. In France, producers and users of glass in the wines and spirits sector have threatened to opt out of Eco-Emballage if their fee is not reduced. For these firms, economies of scale realized through cooperation with firms using other packaging materials are very low because of the existence of alternative networks for the collection of used glass.

Compatibility standards versus conversion technology

In the past, waste compatibility has been ensured by conversion technologies such as landfill and incineration. The waste recovery system was adapted to the waste production system. Under new waste policies, the waste production system must now adapt to the post-consumption system. A compromise must be found between zero compatibility which is very costly and total compatibility which is legally constrained. The choice between the use of a technological gateway or a standard is bargained between industry and government and within firms' consortia.

Environmental efficiency is, in general, lower for conversion technology than for a compatibility standard. The ladder principle embodied in the EC waste policy ranks management techniques in the following hierarchy: reuse, material recycling, physical and chemical treatment (for example, incineration with energy recovery) and landfill. This order corresponds to decreasing compatibility requirements and is assumed to correspond to increasing environmental damage. The preferences of industry and government can diverge. In Germany, for example, the Federal government initially attempted to force industry to develop reuse, whereas the coalition of firms supported cheaper options such as material recycling and incineration.

Bargaining within the coalition of firms regarding the choice of the compatibility instrument, for example, sorting at source versus sorting by waste management firms, is a very sensitive issue. The former leads to the development of single material packaging as a compatibility standard while the latter provides a gateway technology for multimaterial packagings. Glass and paper industries support single material standards, while the plastic industry and composite packaging producers and users are much more reluctant. Recycling techniques and outlets for secondary polymers have been little developed.

Environmental Social Standards

Incentives and information

The environmental arena provides the classic example of standards being used to reduce negative externalities. The incentives for firms to adopt an emission standard are negative since this induces a pollution abatement cost which is not recovered in the marketplace. At a macro-level, environmental compliance does not account for a large share of overall industry costs, constituting approximately 2 per cent of industry's turnover (OECD, 1993b). However, large variations in this average are observed in different industrial sectors (Leveque, 1993a; 1993b). Stringent standards induce particularly high costs for upstream industrial activities. Moreover, there may be large variations within a given industry if different technologies are used by different firms. In this case, information asymmetries may be high and the adoption of environmental standards may be a source of competitive advantage. For example, more stringent environmental requirements for pesticide registration will lead to a significant increase in R&D costs for agrochemical companies, but will also enable larger firms to reduce the market share of companies which specialize in the production of generic pesticides (Nadai, 1994).

The legitimacy of setting or tightening an environmental standard relates to the social benefits of pollution reduction. In general, the valuation and the assessment of these benefits are far from being transparent, especially when the number of pollution victims is large and there are scientific uncertainties. In such cases, firms may challenge the justification of pollution reduction. For instance, they may order controversial studies to demonstrate that the impact of their emissions is negligible, or that the pollution reduction goal is ecologically unwarranted. When moral considerations are universally accepted, or scientific arguments are clearly established, the obstacles which firms place in the path of a new standard are reduced. Government, product users, consumers, citizens or green groups may pose a credible threat to firms and force them to adopt the standard, even if the compliance costs are high. The opposition of firms to the establishment of high environmental social standards increases to the extent at which abatement costs are high, legitimacy is fragile and the industry is homogeneous.

Social standard processes

Unlike compatibility and quality standards, social standards are unlikely to be set through self-regulation. Due to the absence of incentives, coercive power to force firms to reduce pollution is required. In general, coercion comes from government and the standard is set unilaterally.

When a firm expects a competitive gain from the setting of a social standard, it may seek government support in order to force other firms to adopt the same level of pollution reduction. For instance, Mercedes Benz, Haenkel and Du Pont, respectively, have lobbied for the adoption of catalytic converters on cars, limiting phosphates in water and a ban on CFCs. A single firm does not have sufficient coercive power to force its competitor to set the same standard. When an industry is homogeneous, firms may collectively and voluntarily set a social standard only if they expect proactive behaviour to eliminate the risk of a stringent mandatory standard. In the real world, such a case is very unlikely since industry cannot be assured that government will not intervene in the future. Besides, their short-term interest is to wait for the regulation because they will gain a few years without paying the abatement cost. The pattern of self-regulation occurs only when third parties, such as green groups or local communities, have sufficient coercive power. Even in this case, other specific circumstances are required to render government intervention unnecessary. Relevant factors include the level of abatement costs, legitimacy and low transaction costs to permit bargaining, in the Coasian sense, between polluters and pollution victims (Coase, 1937). In industrial areas in Italy, some local pollution damage has been tackled by firms without any administrative intervention.

The dominant pattern for environmental social standards appears to be command-and-control regulation of an industry with high negative private incentives, namely high pollution reduction costs for firms, fragile legitimacy, absence of significant competitive gain and limited technological variation. Some empirical evidence for this comes from industrial and municipal waste management, the energy sector and the chemical industry (Leveque, 1993a; 1993b).

Moreover, there are few theoretical insights to guide policy makers' choices between flexible governance and command-and-control mechanisms in setting environmental social standards. The disadvantages of command-and-control are well known. However, the advantages provided by bargaining and cooperation with industry to set pollution reduction objectives are still unclear (Glachant, 1993). To what extent would bias relating to informational asymmetries (the firm's interest is to exaggerate its pollution abatement cost) and regulatory capture be reduced by flexible governance? To what extent does flexible governance contribute to consensus (Brusco and Cottica, 1993) and the reduction of uncertainties via learning (Van Vliet, 1992)?

Social standards versus economic incentives

Most economists severely criticize the widespread use of social standards instead of economic incentives such as pollution taxes. Standards are viewed as less efficient than economic incentives for two reasons. Firstly, a given pollution reduction objective can, in principle, be achieved at lower cost with price incentives. Secondly, when firms have complied with legal emission standards, they have no further reason to reduce pollution, whereas taxes or subsidies provide dynamic efficiency incentives to improve the environmental performance of processes or products (Bohm and Russell, 1985). However, environmental economics orthodoxy has been criticized (Kemp, 1992) and it can be argued that inadequate attention has been paid to the role which standards may play in reducing transaction costs and exploiting network externalities.

Theoretically, the efficiency issue regarding standards and economic incentives is complex. In a perfect world of complete information, indirect control by prices or direct control by quantities is strictly equivalent in efficiency terms. In a world dominated by informational asymmetries and uncertainties, there is no universal proposition concerning the optimal policy instrument to internalize negative externalities (Weitzman, 1974).

CONCLUSION

This Chapter has shown that the efficiency of environmental policy making relates to both the standards-setting process and the instrument chosen. Such an approach is rare in the literature since, according to the orthodox view, efficiency relates only to the choice of instrument (Bohm and Russell, 1985). It has been demonstrated here that the efficiency of policy instruments depends on the evolution of both informational asymmetries and private incentives during the standards-setting process. More significantly, institutional factors have been integrated into the conceptualization.

Such an approach yields two preliminary sets of findings. Firstly, it casts light on whether there is a single best way to set environmental standards. The analysis of specific factors such as trust in standards-setting institutions, free-rider possibilities or the presence of a credible threat, highlights the circumstances when self-regulation, flexible governance or command-and-control may be the single best way of standardizing. Secondly, it can explain why the same environmental issue (for example, eco-labelling) can be tackled efficiently in different ways in different countries. For instance, limited coercive power on the part of government, or a lower degree of opportunistic behaviour, could make self-regulation a more appropriate method for setting compatibility standards than flexible governance.

12. Changing procedures for environmental standards-setting in the European Community (EC)

Jim Skea

The body of environmental law for which the EC is responsible has expanded rapidly since a coherent environmental policy was first formulated in 1973 (Johnson and Corcelle, 1989). This has mainly been a function of: (a) the growth of concern about environmental issues as experienced in all parts of the Organization for Economic Cooperation and Development (OECD); and (b) a gradual movement of sovereignty in the environmental domain away from the individual member states and towards the EC and its institutions.

This particular context – the impingement of different legal, regulatory and standards-setting practices, coupled with the inevitable strong linkage between environmental measures and questions of impediments to free trade – has lent EC environmental policy-making a distinctive style.

The 'new approach' to harmonization and standards adopted by the EC in 1985 has had comparatively little direct effect on environmental policy, principally because the use of product standards as an instrument of environmental protection has been comparatively rare (CEC, 1985; Task Force Environment, 1990; Haigh and Baldock, 1989). The most prominent example is the case of motor vehicle emission standards which have continued to be discussed in the Council of Ministers (Boehmer-Christiansen and Weidner, 1992).

Nevertheless, there is an environmental counterpart to the 'new approach' in terms of the desire to treat the technical aspects of standards-setting in a different fashion. In the past, environmental standards-setting has been procedurally simple, though often politically more complex. Frequently, standards embodied in the national regulations or laws of influential member states have been translated into EC law and adopted across all member states. The process has been akin to the start-off of a heavy freight train – the environmental locomotive member states have felt themselves pulled back, while those at the rear have felt themselves pulled forwards, the wheels of government squealing.

Present problems stem from three factors:

1. the sheer volume of past and prospective legislation;
2. procedural changes embodied in the Maastricht Treaty which significantly change incentives for regulators; and
3. a wider unease with both traditional instruments of environmental protection and with the processes by which these instruments are developed. As a result, the EC is proposing to make significant changes, both in the selection of instruments and in the way in which responsibilities are shared between public authorities, private enterprise and citizens (CEC, 1992a).

In general terms, this Chapter examines changes in the processes by which environmental standards-setting takes place at EC level. These changes are then illustrated with specific examples drawn from the field of air pollution control.

CHANGING PROCESSES

EC Environmental Policy

The Treaty of Rome which established the EC made no mention of the environment (Krämer, 1990). Nevertheless, the EC has operated an explicit environmental policy since 1973 when it developed the first of its Action Programmes following the 1972 Stockholm Conference on the Human Environment. The first environmental Action Programme operated for a 4-year-period from 1973 to 1976. This was followed by three progressively more ambitious Action Programmes, the last of which ended in 1992.

The EC has now put in place an extensive body of environmental legislation which increasingly serves to determine environmental controls in many of the member states. Table 1 shows that the rate at which environmental measures have been adopted has accelerated. During the Fourth Action Programme, measures were being adopted at a rate of almost 30 per year, almost three times as quickly as in the previous decade (these statistics are based on information in Haigh (1993)). Half of the body of EC environmental law has been put in place since 1986. There is no sign that the rate of rule-making will decline. More than 50 proposed measures await the decision of the Council of Ministers, while a comparable number of proposals are close to completion within the Commission of the EC.

Until the 1987 Single European Act (SEA), EC action on the environment was justified under Article 100 of the original Treaty which permits action which 'directly affect(s) the establishment or functioning of the common market' and the catch-all Article 235 which applies if 'action by the Commu-

Table 12.1   European Community environmental measures

Date	Number of measures
Pre-1973	12
1st Action Programme 1973–76	26
2nd Action Programme 1977–81	55
3rd Action Programme 1982–86	107
4th Action Programme 1987–92	163
Total agreed	363
Proposed to the Council by the Commission	53
Under development by the Commission	50
Total	466

Note: includes Directives, Regulations and Decisions.

Source: based on Haigh (1993).

nity should prove necessary...and this treaty has not provided the necessary powers'. The need to justify environmental legislation in terms of 'the functioning of the common market' gave the EC a powerful tool for action in areas which member states might have regarded as their prerogative. Even before the SEA, product standards (for example vehicle emission standards) could be justified in terms of removing barriers to trade. Process performance standards (for example, power plant emission limits) were also developed on the grounds that different standards in different parts of the EC may give rise to unfair competitive advantages.

Article 130 of the SEA placed EC environmental policy on a firm legal footing. The Maastricht Treaty (the 'Act of Political Union'), ratified in 1993, further strengthens the EC's role. In particular, the cooperation procedure with the European Parliament and qualified majority voting in the Council of Ministers will now be used for most environmental measures. However, three important exceptions are: (a) provisions primarily of a fiscal nature; (b) measures concerning town and country planning, land use and management of water resources; and (c) measures significantly affecting a member state's choice between different energy sources and the general structure of its energy supply. These exceptions relate to areas where, despite the general aspiration to 'ever closer union' expressed in the Maastricht Treaty, several member states jealously guard their sovereignty.

The new procedures will tend to make Community policy 'greener'. During the last decade, many proposed environmental measures have had the support of most member states, with two or three countries forming an effective block-ing coalition. This has led to the abandonment or, most often, the watering down of many measures. In the past, Parliament has typically taken a 'green' position, partly because of the success of Green Parties in European elections, and partly because environmental issues have proved a useful lever for enhanc-ing the status of a fundamentally weak institution. Parliament's new role will certainly promote environmental issues, though the nature of Parliament itself may change as it attracts the attention of more ambitious politicians and a wider range of interest groups, including trade associations.

On the other hand, the complexity of the post-Maastricht cooperation procedures could act as a brake to the adoption of environmental measures. Once the Commission has presented a formal proposal to Parliament and the Council of Ministers, the number of potential obstacles to agreement rises considerably under the new procedure. Problems can arise in the Council itself, in Parliament and its committees, and through differences of view between Council and Parliament.

These potential problems have been recognized by the Commission in its new Programme of Policy and Action in Relation to the Environment and Sustainable Development agreed in early 1993 (CEC, 1992a). This signals a break with the past in terms of both policy articulation and modes of action. In future, the Commission intends to secure as much support as possible for a measure *before* making a formal proposal. The new Programme places greater emphasis on involving the main groups of actors and the principal economic sectors. Although the Programme does not use this terminology, past Com-munity environmental action has largely been based on traditional 'com-mand-and-control' instruments exercised by government over manufacturing industry. There is an aspiration to use a wider range of instruments, notably market-related incentives. Already, measures planned for formal proposal in 1994–96 are the subject of new modes of consultation at both the technical and political level.

The 'Old Approach'

The basic formal procedure for approving environmental measures has so far been simple (Figure 12.1). A Directive is proposed by the Commission, which must then be approved unanimously by the Council of Ministers. The European Parliament and the Economic and Social Committee are consulted, but have enjoyed little power. In the past, the Council of Environment Minis-ters has met twice a year. However, 'informal' Councils have been added and now the meeting rate has effectively moved up to about four times per year.

In developing new proposals, the Commission attempts to operate within the framework of the broad Action Programmes which have operated on a 4–

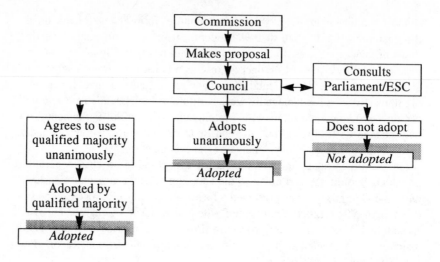

Figure 12.1   European Community environmental decision-making pre-Maastricht

6-year-cycle since 1973. However, high-profile environmental issues such as acid rain and climate change can force rapid adjustments to the agendas set out in the Action Programmes. There is an interplay between the Commission and the Council, and, against a background of high political concern, the Council will often 'invite' the Commission to develop proposals in a specific area.

The Commission's Environment Directorate General (DG-XI), with 150 staff, is severely stretched in terms of the internal resources which it can devote to new environmental measures. It has established several consultative committees which cover specific themes. These committees are made up of experts from member states. In addition, officials rely very heavily on external consultants to provide background information, both on scientific and technical background for new measures and on the state-of-the-art regarding relevant standards or regulations in member states or outside the EC.

In the past, given resource constraints, the Commission has often simply taken standards embodied in the laws and regulations of an individual member state and incorporated them directly into proposed measures. Inevitably, the Commission ends up relying heavily on those member states with the most sophisticated body of environmental law. In practice, Germany has often provided the model for EC measures. Another option for the Commission is to look at environmental laws and standards outside the EC. Notably, the US has provided models for vehicle emissions regulations. While this procedure can lighten the immediate work-load of the Commission, it can create subsequent political problems in the Council. The gap between the most and least ambitious environmental practices of the member states may be very large. The costs of compliance may prove difficult for the less economically developed member states, notably Greece, Spain, Ireland and Portugal.

Another potential problem is that there may be tensions between the instruments proposed by the Commission and the legal or regulatory practices of individual member states (Bennett, 1991). The UK, in particular, found the very legalistic approach to air pollution control embodied in Community proposals difficult to reconcile with its traditional approach which gave pollution control authorities considerable discretion to operate on a case-by-case basis (Haigh, 1984). In practice, the UK has had to make wholesale changes to its style of pollution control (the 1990 Environmental Protection Act gives effect to these changes).

Under the 'old approach', the Council of Ministers has been the key forum for debating new measures. While some environmental measures have been agreed relatively quickly, other more contentious measures have taken years and the final legislation has been very different from the Commission's original proposal. The unanimous decision-making rule has meant that proposed measures have to be broadly acceptable to all member states. When proposals remain on the Council table for a long time, negotiations become 'intergovernmental' in character and the agenda-setting role of Presidency of the Council can become crucial.

Under the 'old approach', it is perfectly possible for the Council of Ministers to get bogged down in discussions over narrow technical matters. The air pollution case study below illustrates this. Mostly, technical issues are resolved to the maximum extent possible through preparatory meetings held between officials or other representatives of the member states. However, there are different approaches to these meetings. For member states with sophisticated environmental rules, these meetings appear purely technical in nature and technical experts are usually sent. Member states that feel threatened by the rules may regard these meetings as political and send trained negotiators with minimal technical expertise. Sometimes remarkably detailed technical issues are allowed to get on to the formal Council agenda. These can prove a very effective 'spoiling' tactic for member states wishing to block or delay a proposal.

Difficult technical issues which have proved to be obstacles to the agreement of a measure have sometimes been ignored or deferred. An example of a technical problem being ignored was the failure to specify uniform measurement techniques in the 1980 Directive on ambient concentrations of sulphur dioxide SO_2 and suspended particulates (Haigh, 1993). The Directive allowed two different, incomparable measurement techniques to be used, reflecting the various practices then current in different member states (this anomaly was removed by a subsequent Directive in 1989). The 1989 Directive on emissions from small cars provides an example of a deferred technical decision (Boehmer-Christiansen and Weidner, 1992). Emission limits for nitrogen oxides (NO_x), hydrocarbons and carbon monoxide (CO) were agreed in terms of grammes over an unspecified test cycle. The subsequent definition of the test cycle, which had important implications for different vehicle manufacturers, was left to technical experts.

The 'New Approach'

The 'new approach' to environmental policy-making is a response partly to the unsatisfactory experiences outlined above, and partly to the more complex formal decision-making processes which the EC will face following the ratification of the Maastricht Treaty. Figure 12.2 shows that the new procedures will involve a complex set of interactions between the Commission, Council and Parliament. The procedures provide for four possible routes for the adoption of a measure and three possible routes for non-adoption. The Council will be able to decide measures on the basis of qualified majority voting, but must work in *cooperation* with the Parliament rather than simply in consultation. (For a qualified majority to be established, 54 of the 76 votes currently available must be cast in favour of a Directive.) The effect of the new procedures has been to move power away from the member states (as represented by the Council of Ministers) and towards the EC's own institutions (Parliament and, to a lesser extent, the Commission).

However, the complex procedures will create many more potential obstacles to the approval of measures proposed by the Commission. The 'new approach' is intended to circumvent these problems through changes in both the *selection* of policy instruments and the *process* through which these instruments are developed. There will be a greater emphasis on the use of instruments other than traditional command-and-control mechanisms, including market-based instruments and 'voluntary' initiatives.

In principle, these should entail less detailed technical direction via EC instruments and more discretion on the part of industry about how to comply with broad aims or respond to economic incentives. (The case of the proposed Community carbon/energy tax shows, however, that business does not always want this type of discretion. Most concerned industries have been asking for 'command-and-control' instruments in this field.) However, there will remain areas where the use of command-and-control instruments embodying standards of a more traditional type will remain. Examples include integrated permitting arrangements for major industrial sites, which shortly will be the subject of a new 'framework' Directive, and air pollution controls.

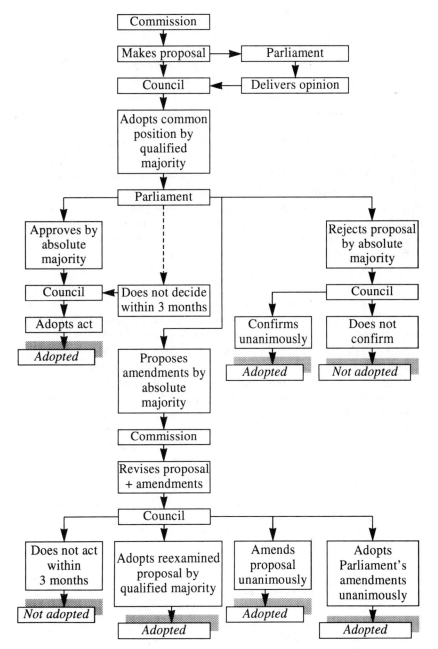

Figure 12.2   European Community environmental decision-making post-Maastricht

The larger innovation may well prove to be the process through which new initiatives are developed. The Commission now intends to seek as broad a support as possible for measures before they are formally proposed to the Council and Parliament. This support is to be gained by involving representatives of industry (and other groups, such as environmental Non-Governmental Organisations (NGOs)) in the process of developing new instruments. In principle, if industry and NGOs are part of the initial regulatory process, they are less likely to attempt to block or modify proposals by approaching sympathetic Directorates of the Commission, by influencing national positions in the Council or by pressing a case in Parliamentary Committees. While the Commission has noted that both industry and wider public interests (as represented by environmental NGOs) should be involved, it is clear that the prime motivation has been to obviate blocking tactics employed by industry. In practice, there is a commonality of interest between the Environment Directorate of the Commission and many of the more sophisticated NGOs.

Consensus building for new proposals will be a time-consuming task. Human resources within DG-XI are limited and a considerably expanded role for external consultants is emerging. While previously the Commission has invited consultants to prepare scientific, technical and economic analyses relating to new proposals, recent calls for tender specify a much deeper role. Consultants are essentially invited to run the entire regulatory/standards-setting process. This includes preparing the underlying analyses, running the consultations with interest groups and writing, in a legally robust form, the draft proposals to be presented to Council and Parliament. These procedures will be heavily reliant on the competence and skills of the consultants used. If this process does not go well then there could be considerable delays in moving ideas out of the Commission in the form of formal proposals.

The new process is at too early a stage to be able to judge its outcome. The following case study illustrates the differences between the 'old' and 'new' approaches for one particular regulatory area.

A CASE STUDY: COMBUSTION PLANT CONTROLS

Overview

The regulation of conventional atmospheric pollution – particulates, SO_2 and NO_X – has a long history in most member states of the EC. This reaches back to the 19th century in some cases. However, the modern era of air pollution control began around 1970 when new techniques for reducing SO_2 emissions were developed following 'technology-forcing' rules mandated under the US Clean Air Act.

Significant Community action in this area did not emerge until the early 1980s when the transboundary nature of acid rain became apparent. In 1976, the EC introduced a Directive on the sulphur content of gas oil with the primary objective of removing barriers to trade. In 1980, a more controversial Directive on ambient air quality was agreed. However, the biggest step forward came with the 'Large Combustion Plant Directive' (LCPD) which limits emissions from power stations and other major industrial sources (CEC, 1988). In US terms, this Directive rolled the requirements of the 1970, 1977 and 1990 Clean Air Act Amendments into a single package. This was proposed in 1983 and agreed 5 years later in 1988.

The present case study looks at the standards-setting processes embodied in the LCPD and at new efforts to expand this regulatory theme by tightening the standards in the Directive itself and extending their scope to cover smaller plant. It should be emphasised that the LCPD represents one of the worst examples of the 'old approach' to environmental regulation. Its proposal, though not its agreement, pre-dates the 'new approach' to harmonization. It was developed in the politically highly charged context of widespread public concern about forest die-back in Germany. The accession of Spain and Portugal to the EC in the middle of the Council negotiations complicated matters. Nevertheless, the case study provides a useful stark contrast between the old and new approaches to environmental measures.

The 'Old' Approach: the Large Combustion Plant Directive

In July 1983, the Federal Republic of Germany (FRG) introduced a new law, the *Grossfeuerungsanlagenverordnung* – Large Combustion Plant Ordinance. This had been prompted by massive public concern about damage to German forests and the suggested link to emissions from power stations, subsequently shown to have been overstated. The ordinance effectively set standards for emissions from new and old plant which had to be complied with within 5 years. The compliance costs have been DM 21 billion.

This section does not address the German story, but it is worth noting that the standards embodied in the ordinance had been conceived much earlier. (The full story in Germany, the Community and the UK is recounted in Boehmer-Christiansen and Skea (1991).) The limit of 400 mg SO_2/m^3 of flue gas had been proposed by the Federal Environment Office, the *Umweltbundesamt* (UBA) as far back as 1978. However, the standards were not adopted because UBA proposed applying them to *existing* plants, while the power industry supported controls applied only to *new* plant. Electric utilities wanted Federal controls on new plant in order to avoid what they saw as arbitrary and excessive demands at the Land level. (For administrative purposes, Germany is divided into a number of states, each called a 'Land'.)

The 400 mg standard was based on a basic judgement about what 'best available' abatement technology, when applied to solid fuel with the quality typically found in Germany, would achieve.

German industry had been adamantly opposed to the new ordinance on the grounds that it would raise power prices and make some sectors (chemicals, aluminium) uncompetitive vis-à-vis producers in other member states of the EC. The German government pressed the EC to take measures which would redress this distortion of trade. Since much acid deposition in Germany originates in other member states, this pressure had both an environmental and an economic rationale.

Within 5 months of the German ordinance taking effect, the Commission had proposed its own comparable measure. For new plants, the quantitative standards proposed were exactly the same as those in the German ordinance. The Commission did not propose plant specific emission standards for existing plants, but instead proposed uniform 60 per cent cuts in emissions by 1995 from all existing plants in each member state. The member states were to have discretion regarding the means for achieving these cuts in aggregate emissions.

This proposal was greeted with outrage in parts of the EC. This was particularly true in the UK where the relatively informal system of pollution control and the absence of orders for new coal-fired power plant meant that emission standards were very lax compared with those achievable by the best available technology. The Commission had acted too quickly and had not prepared its ground. The attempt to translate the law of one member state into a Directive applying to all proved entirely untenable.

The 5-year-negotiation which followed illustrates much that could go wrong under the 'old', top-down approach. Firstly, there was too much ambition in a single package. Obtaining agreement on controls for both new and old plant simultaneously was enormously difficult. This inevitably delayed agreement on standards for new plant which would have been far easier to achieve. The Commission considered splitting the proposal in two, but feared losing its proposal on existing plants completely.

Secondly, the Commission lost control of the process. Having made initial mistakes, the Commission lacked the authority to carry the negotiations through to completion. It was those countries which held the Presidency of the Council between 1986 and 1988 that steered the negotiations to their conclusion.

Thirdly, the standards borrowed from one member state were not easily applicable to others. Pollution abatement equipment had to work at higher efficiencies in Britain to meet German standards because of different coal qualities. Spanish lignite contains so much sulphur that it was necessary to negotiate, over the Council table, a series of new provisions allowing non-compliance with emission standards if performance standards for pollution abatement equipment could be met.

Fourthly, detailed technical issues were allowed to become the subject of high-level political discussion. Remarkably obscure technical issues concerning 'emission limits for combustion plant fired on solid fuel sized between 50 and 100 MW' were discussed by Ministers and were traded off against much more central questions. These technical issues, also addressed by trained negotiators without technical expertise in preparatory meetings, were used to block and delay discussions.

Finally, arbitrary environmental objectives were included. The 60 per cent emission reduction targets for existing plant in every member state could not be justified scientifically. Different energy structures and natural resource availabilities in the member states meant that the levels of effort required to meet ostensibly the same objective were very different.

In the end, the Directive was agreed because the UK found it necessary to remove financial uncertainties surrounding the privatization of its electricity industry. The final measure was very different from the one initially proposed, in that there were many qualifications to the original emission standards and intergovernmental haggling had resulted in a non-uniform set of emission reduction targets for existing plants. Almost no-one's objectives were fulfilled. Emission standards were not entirely uniform and, at the same time, the level of effort expended in each member state has been manifestly different.

New Combustion Controls

The LCPD contained a dynamic clause which obliged the EC to review emission standards for new plant and aggregate emission targets for existing plant in 1994 and 1995, respectively. At the same time, the Commission considers it desirable to extend emissions controls of this type to small plant sized below 50 MW for four reasons:

1. ecological targets ('critical loads') will not be met by current environmental measures;
2. as emissions from large plant decline, smaller plant make a relatively larger contribution to pollution loads;
3. emission standards for smaller plant are set at different levels in different member states – in the case of residual fuel oil, these differences drive trade in refined products; and
4. dirtier fuels forced out of the large combustion plant sector will gradually be switched to smaller plant resulting in lower local air quality through emissions from smaller stacks.

The most highly developed set of emission standards in the EC is in Germany where plant down to 1 MW in size (typical of the commercial or small industrial sectors) is covered by the *Technische Anleitung zur Reinhaltung der Luft (TA Luft)* – technical instructions (air) – developed by the German Society of Engineers (VDI). One option under the 'old approach' would have been to attempt to translate these German standards into EC law. However, it is evident that this is not feasible. Many member states lack the institutional capacity to enforce environmental standards for such small plant. At the same time, the German standards may not be achievable in all member states given the nature of petroleum refining capacity or the quality of indigenous solid fuels. Accordingly, the Commission is using the 'new approach' to work on small combustion plant controls, which will ultimately be decided using the new cooperation procedures with the European Parliament following ratification of the Maastricht Treaty.

Under this approach, it will be necessary to establish consensus on: (a) the bottom end of the size range of plant to be covered by the standards; (b) the form of controls to use – fuel quality standards which would apply to fuel suppliers, or emission limits which would apply to fuel users – compliance and enforcement costs being very different for these two options; and (c) the stringency of the standards to be applied.

The initial stage of this exercise was not noticeably successful. A consultant conducted an initial fact-finding exercise aimed at identifying the characteristics of potentially affected plant in the EC and the various standards employed in different member states (Institut Français de l'Energie, 1992). The views of government departments and trade associations in various member states concerning future standards were sought and a single meeting was held to exchange views. Predictably, a diverging set of views regarding each of the three points listed above was obtained. Ultimately, the consultant recommended a set of emission standards for inclusion in a future Directive to the Commission, but these lacked an articulated rationale of the type which the Commission would need for the explanatory memorandum which it customarily sends to the Council of Ministers.

This task has now been retendered along with work on revisions to the Large Combustion Plant Directive. The call for tenders emphasizes the very wide range of qualifications which will be needed if external consultants are to help the Commission to put new environmental measures successfully on the Council table. These will include a sound technical and economic grasp of the subject matter, thorough knowledge of the EC's operations and the ability to draft legal documents. Inevitably, consortia of different organizations are forming to put in tenders. Although the 'new approach' may avoid difficult formal procedures after a measure has been proposed, the fundamental issues and divergences of interest which must be resolved in formulating any new standard must be addressed in some forum. It remains to be seen how successful this approach will be, given the wide range of players and interests which will be involved in the initial stages of the new regulatory process.

CONCLUSION

EC environmental law has expanded rapidly over the last 20 years and will continue to grow. As with processes for the harmonization of technical standards more generally, a 'new approach' to environmental rule-making in the EC is emerging. The motivations are to remove detailed technical discussions from the Council table, to involve the main groups of actors (governments, industry, the public) in the process and to avoid delays to legislative agreement following the application of qualified majority voting and the cooperation procedure to environmental measures. The 'new approach' involves a conscious move away from traditional instruments (usually standards-based) towards market-based instruments and 'voluntary initiatives'.

It is too early to say whether the 'new approach' will yield the hoped for benefits in terms of consensus and the smoother approval of environmental measures. However, the example drawn from the air pollution case study shows that real changes in procedure have taken place. Nevertheless, there are limits to how far market-based instruments can replace more traditional command-and-control approaches. There must be some forum for resolving difficult technical issues which have significant implications for the competitive advantage of both member states and companies. While the 'new approach' may have found a better location for this type of discussion, much will depend on the capacity of the Commission and its advisers to steer the process, taking account of the ambitious programme of action which has been put in place.

13. The standards processes in telecommunication and information technology

Stanley M. Besen

This Chapter analyses the processes by which compatibility standards are established in the telecommunication and information technology industries. It addresses two related questions:

1. What determines the processes that are employed to set standards?
2. How do the processes that are used affect whether a standard will be chosen, how long it will take to establish a standard and what that standard will be?

Note that the way these questions are posed indicates that both the nature of the standards process and its outcomes are endogenous and affected by factors such as an industry's market structure, the nature of the products or services it supplies and the types of standards that are employed. Thus, whether an industry will seek to improve the process by which it chooses standards, or allow an apparently inefficient standards process to continue, are things to be explained.

The Chapter focuses primarily on the private voluntary standards process – sometimes referred to as cooperative standards-setting – as opposed to non-cooperative or *de facto* standards-setting which takes place in the market, or *de jure* standards which are set by government. Thus, it is concerned with: (a) how many participants are allowed to participate in private standards bodies; (b) who these participants are; (c) how resources to support standardization activities are obtained and organized; (d) how decisions are made as to what will be standardized and what the standard will be; and (e) how many standards organizations there will be.

Since the 'default' mechanism is *de facto* standardization through the market, one objective of this Chapter is to explain both why and when an industry may choose to use private voluntary standards-setting organizations and what form the standards process will take. It examines, for example, why

136

one industry may choose a slow and cumbersome voluntary standards process with the result that most, if not all, standards are set through the market, while another adopts a mechanism that leads to rapid agreement, or at least rapid decision, so that voluntary standardization becomes the primary method through which standards are adopted. Similarly, it explores why the organization of standards activities in an industry may change over time.

WHY ORGANIZE STANDARDS-SETTING?

To those who participate in the formal standards process, it may seem strange to ask 'Why is there a need for formal standardization?' After all, no standards body chose VHS as the videocassette standard, or MS-DOS as the standard personal computer operating system, or the Intel chip as the industry standard micro-processor.

One answer is that competition to become the *de facto* standard may ensue for many years before the network of one sponsor grows to the point at which it achieves dominance. In the interim, the growth of the market may be adversely affected as consumers delay purchases while waiting to see which standard is the victor (Besen and Johnson, 1986). By organizing standardization centrally, and thus eliminating uncertainty about the identity of the winning standard, the delay in market development may be reduced.

A second reason is that centralized standards-setting may avoid leaving some users as 'orphans' – stranded with a non-standard technology (Besen, 1992). If products are produced and marketed only after a standard has been adopted, a situation in which users are forced either to remain in very small networks or to incur switching costs may be avoided.

Finally, the economic theory of standardization teaches that the market process can adopt the 'wrong' technology as the standard. This can occur because adopters do not take into account the effects of their choices on the well-being of others, that is they generate network externalities, so that the uncoordinated process can produce an inefficient standard. Moreover, the strategic behaviour of sponsors can produce either less than the efficient amount of standardization, or lead buyers to the wrong choices. In principle, at least, a central decision-maker could offset these inefficiencies by internalizing the network externalities.

Of course, just as market failure does not, by itself, justify government intervention, the fact that the *de facto* standards process is less than perfect does not necessarily mean that the private voluntary standards process will perform better. To determine whether that is the case, it is necessary to examine how cooperative standards-setting actually works and to compare its performance to that of the market.

SPONSOR OBJECTIVES AND THE CHOICE OF STANDARDS PROCESSES

Sponsors of technologies may prefer the same standard, different standards or no standard at all (Katz and Shapiro, 1985, 1986; Farrell and Saloner, 1986). No less obvious is that sponsors may be concerned about the process that is used to select standards. Their views regarding the choice of process to employ are related to the outcomes they favour.

Four basic situations are possible.[1] Firstly, all sponsors may have common interests in the standard that is selected, so that they all prefer a mechanism that quickly and efficiently 'discovers' which standard best serves those interests. This might be thought of as the 'technocratic' view, where participants selflessly attempt to serve the common good. Here, one might expect the process to be open to all, since widespread membership need not make decision-making more difficult. Additional members bring added resources to support the process and acceptance of the standard is enhanced if more participants are involved in its adoption.[2] Moreover, because interests are congruent, decision-making can be by consensus.

Secondly, sponsors' interests may be directly opposed, in that each prefers that others should adopt its standard, but each would prefer no standard as an alternative. This outcome is most likely when the benefits to the industry as a whole from having a standard are small relative to the differential gain obtained by the winner of a standards battle. Thus, for example, if market growth is not inhibited as the competition to become the standard ensues, while the winner of a contest to become the *de facto* standard stands to obtain a very large share of the resulting gains, and competition 'within' a standard dissipates a large share of industry profits, sponsors are unlikely to turn standards-setting over to an organization that can resolve conflicts quickly.

The same outcome can also occur where users have widely divergent preferences, so that market growth is stimulated as much by variety as it is by compatibility. As a result, some firms may prefer to dominate particular market niches rather than compete within a generally accepted industry standard, making agreement on a standard difficult.

In this case, the formal standards process provides only one of the fora in which standards battles occur, and sponsors prefer a process in which it is easy to commit to one's own standard and difficult to concede to another. They will also oppose the adoption by standards organizations of rules that make it easy to break deadlocks, or that shift power to non-sponsors who are likely to care less about which standard is adopted than that a standard is chosen. Farrell observes that the rules chosen for the selection of a standard will be more likely to affect the speed of the process adversely if product vendors would prefer not to have a standard at all (Farrell, 1993). Stalemates are likely to occur in these circumstances, not because sponsors have chosen the wrong rules, but because that is what sponsors prefer. The result is that standards-setting is usually shifted to the market.

Thirdly, each sponsor may prefer its own standard, but each would prefer another standard to none at all. This case is most likely when the gains from having a standard are very important – perhaps because consumers will delay their purchases if they are unsure about which standard will ultimately emerge – and the gains from having one's own technology chosen as a standard are relatively small. Here, the standards process primarily provides a vehicle for effecting compromise (Besen and Farrell, 1994).

Since sponsor interests are somewhat divergent, the process will allow for some competition, but it will also encourage agreement.[3] For example, the standards rules may provide for extensive sharing of intellectual property, so as to increase the gains from concession.[4] In addition, they may include rules for voting when consensus cannot otherwise be reached.

At the same time, sponsors may wish to limit membership in standards organizations both to prevent an extended period of competition and to increase the chances that their standard will be selected. However, the membership must be large enough to ensure that any standard will be accepted by non-members. In addition, users may be welcomed as members if their presence is useful in breaking deadlocks.

A variant of this case occurs where there is a single dominant sponsor that can dictate the standard, although smaller players would prefer different ones. The smaller players concede here because they are better off having the dominant firm's standard than none at all, and they have no reasonable expectation of having their own technologies being adopted as the standard. Moreover, the adoption of standards can be quite rapid since the pace is determined entirely by the dominant firm (Farrell, 1993). Here, the precise rules employed by the standards organization are likely to be of little importance, since everyone recognizes the dominance of a single sponsor.[5]

These rules are, however, likely to become important if the dominant firm becomes less powerful. How the rules will change will depend on whether, in the new situation, firms wish to engage in a standards battle or prefer to concede if they cannot win. This, in turn, depends on how reaching agreement affects the growth of the market and on the advantage to a firm of having its technology become the standard.

Finally, at least one sponsor may prefer that others be unable to produce compatible products, in which case its participation in the standards process, if it occurs at all, reflects a desire to slow the process rather than to make it work. This differs from the previous case since here the dominant firm wishes to prevent others from joining its network and there it wished to encourage them to do so.

This outcome will occur where the growth of the market that results from an open standard is small relative to the increase in competition to which the dominant firm is subject if smaller firms can achieve compatibility. Here, there will be conflict about the standards process, with the dominant firm preferring a slow, cumbersome process and others preferring one that is open and gives them easy access to the dominant firm's technology.

TELECOMMUNICATION AND INFORMATION TECHNOLOGY STANDARDIZATION

The aggregate benefits obtained by sponsors from standardization can be large both because the presence of standards promotes market growth and because competition to determine the standard dissipates a large share of industry profits. Moreover, standardization is attractive to sponsors when the differences between the rewards to winning and losing sponsors are small. In such cases, there are large gains to all sponsors from quick agreement on a standard and few benefits from delay. As a result, all parties will want a standards process that quickly finds a suitable standard and does so at reasonable cost.

By contrast, if standardization produces lower industry profits because intensive price competition drives prices close to cost, and if a winning sponsor obtains a very large share of industry profits, it will be more difficult and time consuming to achieve coordination. And, as a corollary, standards organizations can be expected to be governed by rules that make it difficult to achieve agreement.

Traditional telecommunication standards-setting provides an example of a situation in which the pressure on the parties to agree on standards was significant. Where network externalities are very great, consumers are likely to defer choices until they can be certain that they will not be stranded with a non-standard technology. In most countries, a single dominant provider could determine standards with great assurance that others would follow. In addition, industry participants were often not direct competitors, which further weakened any conflicts that might arise among them. For example, the independent telephone companies in the US did not compete directly with AT&T, so that they would not be placed at a competitive disadvantage by adopting standards established by the Bell System. Similarly, international telecommunication standards were determined largely through the cooperation of geographically separated national monopolies. In these circumstances, the nature of the standards process was likely to be relatively uncontroversial.[6]

In the mainframe computer industry, the situation was somewhat different. A single firm, IBM, controlled and set standards for a large portion of the industry, but industry participants were direct competitors. Nonetheless, some competitors did not view compatibility with standards adopted by the dominant firm as necessary to market success and some firms explicitly chose standards different from those adopted by IBM. At the same time, other firms did prefer to adopt IBM standards and, indeed, often complained that IBM made it difficult for them to do so. In these circumstances, conflicts over the industry standards process were primarily about access to IBM standards, rather than about the way in which IBM standards were adopted.[7]

Clearly, the situations in both industries have changed dramatically in the last two decades. In telecommunication, to varying degrees, providers of terminal equipment, interexchange service and enhanced telecommunication services, have begun to compete with the traditional monopoly carriers. Moreover, the market growth in the variety of telecommunication services offered means that, although compatibility remains important, the required degree of compatibility may be limited and significant variety in offerings is possible without constraining market growth. Indeed, variety itself may be an important source of that growth.

With the introduction of competition, the standards process in telecommunication was almost certain to have become more contentious as new entrants challenged the hegemony of the incumbent monopolists in determining standards. As in the mainframe computer industry, the issue of access to the standards of the dominant carriers has become more important.[8] The process has also grown more fragmented since success in the provision of many services no longer requires adherence to a single set of standards and the number of market participants, both suppliers and users, has grown rapidly.

In reaction to these developments, new telecommunication standards bodies have come into being in many countries. A fundamental objective of these organizations is to avoid the controversies and delays that are likely to accompany the developments described here. However, although these organizations may ameliorate such problems somewhat, they are likely to be only partially successful in doing so. The reason, of course, is that the problems cannot be eliminated solely by changes in procedure.

The changes in the computer industry have some parallels to those in telecommunication – an enormous increase in the number of firms and the decline in the role of a dominant firm in setting standards. At the same time, however, other firms – notably Microsoft in operating systems and Intel in micro-processors – have, to some degree at least, taken on the role formerly occupied by IBM. Some firms have apparently been driven to agree on rival standards in order to compete more effectively with Microsoft, but many others have opted to join the Microsoft bandwagon (Johnson, 1993; Fisher, 1993; Clark, 1993).

Other changes probably have increased the importance of standards, and thus made the process of developing standards more contentious. In particular, the growth of the personal computer market has made issues of compatibility more important. It may no longer be possible to serve a significant market niche with hardware or software that does not conform to an industry standard (Yamada, 1993). If that is the case, it becomes more important either to win the standards battle or adjust to the winner's standard.[9] And, as a corollary, winners may try to make it difficult for the losers to conform to their standards (Besen and Farrell, 1994).

In this new environment, there are competing forces. On the one hand, market growth is now probably more dependent on the development of standards than it was in the past. On the other, standardization is likely to be more difficult both because of the increase in the number of industry participants and the increased stakes that each has in the outcome of the process. Despite the greater importance of standardization, or perhaps because of it, more standardization may be shifted to the *de facto* process as cooperative standards-setting proves incapable of dealing with the additional strains placed upon it. Alternatively, the result may be the Balkanization of standards-setting, with different groups coalescing around different standards.

HOW TO ACCELERATE THE FORMAL STANDARDS PROCESS

Most criticisms of the formal standards process focus on the slow pace at which standards are adopted, but complaints are also voiced about the standards that emerge from the process themselves. To some degree, these goals are in conflict. If there were no concern about the quality of a standard, decisions could be made quite rapidly. But many believe that the process could be accelerated without significant sacrifice in quality.

Several ways have been proposed to achieve this objective. Firstly, because achieving consensus is often time consuming, a number of organizations have adopted voting systems for use when consensus cannot be achieved (Besen and Farrell, 1991). These systems usually require substantially more than a simple majority before a standard can be adopted, and voting generally occurs only as a last resort, but the presence of these rules can be important in breaking a deadlock. Indeed, the mere presence of the voting alternative can affect behaviour, even if it is rarely employed.

Secondly, standardization can be begun prior to the time at which the large investments that have been made by sponsors make subsequent agreements more difficult. Starting early does not guarantee consensus. but it does increase the likelihood that consensus, will be achieved (Lehr, 1993).

Thirdly, the standards process can provide for less disparate rewards between winning and losing sponsors. If, for example, the winner is required to offer low cost access to the patents embodied in the standard to losers, resistance to adoption can be reduced. The winner may also agree to share future developments with losers on a timely basis, or even to permit them to participate in the development process.

Finally, the process can be expedited by situating standards development work in the standards bodies themselves. This tack has been taken both by the European Telecommunications Standards Institute (ETSI) and the Open Software Foundation (OSF), both of which have their own standards development personnel, using either regular employees or assignees from members' organizations (Besen, 1990; Saloner, 1990).

Undoubtedly, the adoption of these changes will make the voluntary standards-setting process more effective than it otherwise would have been. At the same time, it is no contradiction to conclude that the process will prove less satisfactory than it had been previously. Establishing a given standard now requires either additional time and resources (or both) because of the increase in the number of participants with conflicting interests and, perhaps, because of an increase in rewards to the winning sponsor. In response, the rules used to adopt standards are changed in an attempt to reduce the cost and time required to reach agreement on a standard. Because the underlying conflicts are greater than they previously had been, the process takes longer than it did previously, although not as long as it would have taken had the reforms not been adopted.

WHY ARE 'IMPROVEMENTS' SO DIFFICULT?

Concerns are frequently expressed that the standards process works too slowly, and claims are made that every effort is being made to make it work more rapidly. For example, the Secretary-General of the International Organization for Standardization (ISO) has stated that 'ISO and IEC have demonstrated their willingness to adapt their procedures, and the criteria for participation in their work, to insure that ISO/IEC IT standards are produced in the most rapid possible times scales...' (Eicher, 1990: p. 39).

Similarly, the International Telecommunication Union's International Consultative Committee on Telegraphy and Telephony (CCITT) Modified Resolution No. 2 developed 'an accelerated procedure to be used for the approval of new and revised Recommendations between Plenary Assemblies'.[10] Despite such efforts, improvements have been difficult to make and dissatisfaction continues. Some of this is undoubtedly due to the fact that standards-setting is inherently difficult and time-consuming, but some is also due to the inherent conflicts with which the process usually must deal. Otherwise, it is difficult to see why, if apparently everyone wants the process to work more rapidly, there is so much difficulty in bringing change about.[11]

A corollary is that we should expect progress in reforming standardization either in those situations where interests are congruent, or where the benefits of being on a large network are large relative to the gains from having one's own technology become the standard. Where reforms are attempted in other situations, they are likely to be of only limited success.

WHY DO STANDARDS BODIES PROLIFERATE?

It has often been remarked that, in response to a process that is perceived to be too slow, the number and variety of standards bodies is increasing rapidly. If one set of industry participants is willing to compromise on a standard but others are not, the former can be expected to organize separately to avoid some of the costs of coordination and to accelerate agreement on a standard (Shankar, 1992; Reynolds, 1990). Nonetheless, uncoordinated growth in the number of standards bodies may eventually contribute to the slowness of the process. Indeed, one commentator has referred to the proliferation of such bodies as 'alarming' (Rankine, 1990).

Two difficulties result when new standards organizations are established in response to the perceived failure of existing organizations to develop standards in a timely manner. Firstly, the creation of new organizations diverts resources from existing standards bodies. If, as seems likely, economies result from developing standards in a coordinated manner, and the amount of resources available to support the process is fixed, an increase in the number of bodies reduces the amount of resources that are available to develop any single standard. At the other extreme, achieving the development of the same number of standards without adversely affecting their 'quality' requires an increase in the amount of resources that are devoted to the standards process.

Secondly, an increase in the number of standards bodies may lead to an increase in coordination costs as standards that are developed by different organizations themselves must be brought into conformity. Although a unitary standards body must also effect such coordination, the cost of doing so is likely to increase as the number of standards bodies increases.

Despite the problems created by the proliferation of standards bodies, however, it should be noted that new organizations can produce significant benefits. In particular, actual or potential competition among standards bodies can, in some circumstances, actually accelerate the pace of standards development as one body attempts to prevent others from displacing it (Besen and Farrell, 1991). In any event, preventing proliferation is likely to prove difficult if some participants become dissatisfied with the performance of an existing standards body.[12] Moreover, the expansion in the number of standards organizations is likely to continue even if its net effect is to make the overall standards process less efficient.

CONCLUSION

The moral of this Chapter is that one cannot analyse the performance of the standards process without understanding what the participants in the process expect to obtain from it. If participants prefer no industry standard, or will settle only for their own, agreements on a standard are unlikely and changes in the process to make agreements easier to achieve will be difficult to bring about. In short, improvements in the process will occur only if participants believe they are better off with those improvements.

This does not mean that changes in the procedures used to produce standards are unimportant. Indeed, they are likely to be most important precisely where changes in industry conditions have shifted so that the benefits from agreeing on a standard have become more important relative to the potential gains from competing to become the *de facto* standard. In such cases, the greater importance of standards provides the impetus for the improvements in the standards process. Attempts to improve the process are most likely to be effective if all industry participants would prefer to compete within an industry standard rather than compete to determine what that standard should be.

NOTES

1. These distinctions are drawn from Besen and Farrell (1994) where they are used to analyse the (primarily) marketplace 'tactics' that sponsors use in each of these situations.
2. There may, however, be conflicts about who will support the process. Indeed, if the process serves the common good, there may be little reason for anyone to support it, especially if the number of beneficiaries is large (Weiss and Toyofuku, 1993). On the other hand, free riding may be restrained because: (a) the benefits a firm obtains from using a widely available technology are reduced if it has not participated in the development process (Rosenberg, 1990); and (b) there is an incentive to share information when the interests of firms are congruent (Lehr, 1993).
3. Farrell (1993) emphasizes the bargaining aspects of such processes because he assumes there are no side payments, but he also assumes that winning sponsors must make their technologies available to others on reasonable terms.
4. Reynolds (1990, p. 438) observes that '...vendors should create a mechanism to compensate each other over the selection of a standard, where there are clear winners and losers over the selection'. What he does not note is that establishing compensation mechanisms may be difficult if sponsors would rather 'fight' than 'switch'. Farrell (1993) examines whether requiring loose intellectual property protection adversely affects the incentive to improve technologies and concludes that the gains from accelerating the adoption of a standard may more than offset the fact that others share in the resulting gains.

5. If, however, the small firms fear eventual abandonment by the dominant firm, they may attempt to form their own standards organization. This apparently occurred when a number of software vendors formed the Open Software Foundation because they feared that AT&T might close the Unix standard to them.

6. This is not to say that they were completely uncontroversial, since the costs incurred by a provider were affected by the standard chosen. The point here is that one source of conflict, direct competition among sponsors in selling to final consumers, was absent.

7. There were, however, controversies about the adoption of non-IBM standards by those firms that were not part of the IBM network as well as complaints that IBM behaved unfairly in promoting its products and standards.

8. For example, under Federal Communications Commission rules [47 CFR 64.702], communication common carriers are required to provide information about changes in network design and technical standards to rival suppliers of customer-premises equipment and enhanced telecommunication services before the changes are introduced into their networks.

9. IBM and Apple have begun the joint development of their own high end graphical workstation and operating system which would compete with hardware using Windows. This arrangement seems to reflect both IBM's weakness, which requires that it seek partners if its standards are to be successful, and Apple's dissatisfaction with its continuing role as a niche player (*Information Week* (1993) p. 50).

10. The text cited by Wallenstein (1990: pp. 229–32) is from the modified CCITT Resolution No. 2, IXth Plenary Assembly, Melbourne, 14–25 November 1988.

11. A complementary explanation concerns the difficulty standards bodies have in obtaining resources to carry out their work. Even if there were no inherent disagreements regarding the choice of standards, sponsors still might not agree on the importance of having a standard and, thus, how to divide the cost of conducting the needed work. For a taxonomy in which these two sources of conflict are discussed see Besen and Saloner (1989).

12. Besen and Farrell (1991) argue that the International Telecommunication Union (ITU) may be unable to resist a shift in the locus of power in setting international standards to a consortium of regional standards organizations.

14. Standards-making as technological diplomacy: assessing objectives and methodologies in standards institutions

Richard W. Hawkins

The practice of negotiating voluntary technical standards in committees under the auspices of designated standards-making institutions is barely a century old. The historical antecedents stem partly from economic phenomena which accompanied the maturation of industrial processes of mass production and distribution – chiefly associated with the increasing decentralization of production and distribution capacity (Thompson, 1954; Hemenway, 1975). They also stem from the institutionalization of trade and professional skills. Increasing technical specialization conferred a measure of authority upon technicians in specific industries to define standards governing many aspects of the design, production and exchange of technologies (Sinclair, 1969).

Since the turn of the century, a reasonably coherent set of principles has evolved to govern the process of standards-making in committees. These principles are of a *quasi* juridical nature and seek primarily to assure that standards committees are: (a) made up of a representative cross-section of 'stakeholders'; (b) afford 'due process' to each committee member; (c) allow for public scrutiny and comment; and (d) produce technical documents which reflect a consensus of expert opinion. The entire process is usually considered to be 'voluntary' – committee members are nominally 'volunteers', and implementation of the documents produced is normally not compulsory.

The concepts of consensus and voluntarism have largely defined the institutionalized standards process. Nevertheless, both are now frequently challenged by industry, the policy-making community, public opinion and even by standardization organizations themselves. Voluntary/consensus standards-making systems are now increasingly juxtaposed with social and industrial contexts that, arguably, were not envisaged when the methodologies of national and international standards-making organizations were initially formulated.

Standards are now often far more than mere *ex post* consolidation exercises of existing technical practices. The speed at which technology has developed during the past 40 years, coupled with the 'network' and 'complementarity' effects visible in many technical systems has demanded in many cases that standardization issues be discussed at the planning and design stages of new technological initiatives.

Also, in recent years, increasing public awareness of the role of standards in industrial production and distribution processes has resulted in an unprecedented number of public policy initiatives. Some of these seek merely to adjust the process in the hope that it may become more responsive to a greater range of social and technical criteria. However, standardization can also be used in its own right as a policy instrument. Standards can be used to promote either the liberalization or the protection of domestic markets. They can also be pressed into service to coordinate aspects of the innovation process at national and transnational levels according to politically specified priorities.

The basic question is: 'Are the traditional structures and procedures of existing standards institutions appropriate to the changing roles of standards in contemporary societies?' A look at some of the practical and theoretical perspectives will indicate that the answer cannot be a simple 'yes' or 'no'. An evaluation of methodological principles for committee-made standards must follow to a large extent from a re-evaluation of the purpose of standards institutions in rapidly evolving social, commercial and technical circumstances.

THE CONSENSUS IDEAL AND THE NEGOTIATION OF TECHNICAL AGREEMENTS

The root cause of the general acceptance of the ideal of consensus in standards-making may well lie in the genealogy of standards institutions. The first real prototypes of modern standardization organizations were the various international trade and technical congresses which were established in the mid- to late 19th century (Adams, 1956; Coonley, 1956). Typically they were tied to national diplomatic agenda, especially where these concerned international trade or the control and coordination of communication systems. Organizations like the International Telegraphic Union – the forerunner of today's International Telecommunication Union (ITU) – and the Universal Postal Union, and the International Bureau of Weights and Measures, were products of this era.

Although recent attempts have been made to reorganize the relationship between public administrations and private firms in the ITU, standards-making in that organization has a long history of being closely linked with its diplomatic and governance functions (Cowhey, 1990; Codding, 1991; ITU, 1991). Even where the early institutional standardization initiatives came from outside the public sector, there remained strong overtones of 'diplomacy' as it would have been understood in contemporaneous intergovernmental negotiations. The first major agenda items confronting the forerunners of today's International Organization for Standardization (ISO) and International Electrotechnical Commission (IEC) concerned materials testing and units of measurement, respectively (ISO, 1986). This agenda was closely related to several parallel intergovernmental initiatives on trade.

Indeed, many insights into the actual role of consensus and voluntarism in standards-making can be obtained directly from theories of political negotiation. It is frequently professed, for example, by standards-makers and academic analysts alike, that consensus is obtained through the political arts of concession and compromise. This is the root of a common anecdotal definition of standards – that they are solutions mutually unacceptable to everyone!

Where they concern technology, however, notions of compromise and concession can be fundamentally troublesome if they are not linked to precise identification of those particular aspects of the negotiations which may offer scope for elasticity of response. In negotiating technical agreements, both terms imply that technological alternatives are actually available to the negotiators – a condition that may not apply. The ability to compromise technically may be seriously limited by factors ranging from existing levels of commercial commitment to selected technologies, to limitations in scientific understanding.

It is instructive to examine the roots of 'consensus' as a term in the social sciences before looking at it in the context of standards-making. Partridge ascribes the academic origin of the term to 19th century sociologists. He notes that for them '...the notion of consensus is not limited in its reference to agreement about beliefs, attitudes, values, norms, objectives, etc., but is used much more widely to refer to the interdependence or inter-connectedness of the parts of a society – to what we might rather think of as the systemic character of societies' (Partridge, 1971: p. 75). Partridge notes further that the earliest definitions of consensus in social theory relate the concept of 'agreement' to a 'voluntaristic' context of social organization (Partridge, 1971).

From the standpoint of political theory, Zartman (1977) notes three modes of decision-making: (a) judication, by which conflict is resolved by a single authority; (b) coalition, by which choices are made by 'numerical aggregation'; and (c) negotiation, by which the participants '...come off better in the agreement than in the absence of the agreement' (Zartman, 1977: pp. 69–71). In the present context, it should be noted that standards are established by all three of these mechanisms and that a different type of consensus-building can apply in each of them.

Judication often involves consensus-building in designated bodies of authority – consensus which can be tempered by information solicited from external parties. Coalition typically yields a situation in which one side is simply overwhelmed numerically by another. Thus, individual consensus groupings can be placed in opposing positions. Negotiation implies the formation of a consensus among the whole negotiating group, notwithstanding that various forms of coercion can still come into play during a negotiation process.

Zartman also chronicles several approaches to the study of negotiation, all of which fail, for different reasons, to take account of the practical situation in which negotiators find themselves. Examination of the behavioural and motivational psychology of individual actors is not adequate to explain institutional factors. Many forms of economic analysis are dependent upon artificial constructs and theories of optimality which have demonstrated little power to predict outcomes when applied to actual negotiating situations. Likewise, strategic analysis employing game theory can become prisoner to assumptions concerning 'rational-choice behaviour' which may or may not be present in actual negotiating situations in any consistent way. Process analysis – particularly the study of 'concession-making' – has strengths in that it recognizes the changeable nature of negotiation processes while they are in progress, but it too can be at odds with the way negotiations actually take place (Zartman, 1977).

Every analysis of negotiation must confront the possibility that under some circumstances parties to the negotiations will accept less than optimum outcomes. This can obviously be encouraged if compensation in some form is paid to disadvantaged parties in return for their agreement. However, there may also be circumstances in which agreement itself becomes more important than its terms.

Buchanan and Tullock (1962) set the concepts of individual and collective decision-making processes in opposition, insisting that choices between the two methods are made on the basis of their respective internal and external costs to the decision-makers according to individual circumstances. However, in terms of formal logical analysis, Lehrer and Wagner (1981: p. 13) propose that there may also be circumstances in which the logic of pursuing a collective understanding is inherent: '...the information individuals possess ... may rationally commit them to consensus. ...the course that amalgamates the available information is more reasonable than one that neglects it'.

All of these comments have a certain resonance when compared with existing studies of standards. From the perspective of the standards developer, for example, Verman (1973) has maintained that by committing themselves voluntarily to the development of a standard by consensus under mutually and publicly agreed procedures, actors might actually be applying pressure upon themselves to implement the standard.

In their game-theoretical model of the outcome of standardization committee processes, Farrell and Saloner (1988) limit themselves to modelling choices between fixed alternatives. Nevertheless, they admit to the probability of compromise, and they also show that they are fully aware that idiosyncratic negotiation factors which are unrelated to the technical and/or commercial alternatives may bias the process away from the most 'rational' outcome. In their empirical study of standards committee outcomes, moreover, Weiss and Sirbu (1990: p. 132) concluded that factors other than the objective assessment of technology were 'significant predictors of the possibility of adoption'. *Ex post* analysis of committee outcomes by individual participants revealed that most retained the view that the solutions they themselves had proffered remained the 'best' solutions irrespective of the eventual decision of the committee (Weiss and Sirbu, 1990).

Zartman (1977: p. 78) maintains that negotiation is actually a far more prosaic activity than theorists often suppose – that negotiations only come to a conclusion on the basis of the discovery of the 'formula and implementing detail' which will satisfy the objectives of the negotiating parties. 'Above all, negotiators seek a general definition of the items under discussion, conceived and grouped in such a way as to be susceptible of joint agreement under a common notion of justice. Once agreement on a formula is achieved, it is possible to turn to the specifics of items and to exchange proposals, concessions, and agreements.'

Intriguing and useful as many of the above theories might be, it is nonetheless constructive to consider the practicalities of standardization committees in the light of Zartman's admittedly pragmatic criterion for the negotiation of positive outcomes. In so doing, we are forced to expand our scope when considering the actual and potential uses of committee processes. The basis also emerges for a constructive critique of the institutionalized standardization structure and related public policies.

CONTEMPORARY STANDARDS INSTITUTIONS AND THE CONSENSUS IDEAL

In 1990, ISO/IEC published the results of an extensive survey of projected demand for international standardization (ISO/IEC, 1990). In many ways, however, the results of the study were less illuminating than some of the assumptions which underpinned it. The study proceeded from a general observation (not disputed here) that standards are now often closely linked to Research and Development (R&D) and that a measure of standardization is often an essential precondition for establishing markets for new technologies. The report detected a problem with the consensus concept given the difficulty in identifying common interests between users and suppliers where the technology concerned was rapidly emerging but not yet implementable (ISO/IEC, 1990).

The report proposed that the solution to this dilemma was for ISO/IEC to engage in some form of preliminary or *quasi* standardization. This would involve direct participation in special committees at the international level by individual firms involved in developing new technologies for which high demand could reasonably be anticipated provided that the implementation environment was standardized to some *a priori* degree (ISO/IEC, 1990). Although framed in terms of increased responsiveness and efficiency, this proposal nevertheless implies two significant structural changes.

Firstly, it effectively bypasses an entire level of consensus-building. Since their foundation, ISO and IEC have been constitutionally and philosophically committed to seeking international consensus on the understanding that the inputs of each of their national member bodies would reflect a prior determination of consensus between 'stakeholders' at the national level. This was always potentially a scenario favouring coalition over negotiation, but the direct injection of 'free-agents' – particularly in the form of multinational companies – would complicate this scenario significantly. Secondly, the proposal effectively defines ISO and IEC as potential competitors to other standards bodies. In a real sense, eliminating duplication of effort at national and international levels could have the practical effect of centralizing the primary initiative to standardize at international rather than localized levels.

In anticipating industry and user responses to proposals for *quasi* standardization initiatives, it is instructive to compare the background assumptions of the ISO/IEC study with the experience of, arguably, the most significant forward planning initiative taken by these organizations to date – the Open Systems Interconnection (OSI) Reference Model.

OSI is an attempt to establish an international framework for the development of computer interconnection standards in support of a decentralized multi-vendor data processing environment. OSI is supported by a wide range of public and private sector institutions having substantial computer procurement requirements. It remains the basis of government policies for procurement and systems planning in many countries.

Over the years, however, OSI has shown itself to be more 'figurehead' than 'vessel'. Although remaining the focal point of the ideal of the multi-vendor environment – perhaps akin to the 'formula' in Zartman's terminology – the application of OSI standards to actual products in the market has been conspicuously limited. Solving individual interconnection and interoperation problems has become largely the province of proprietary network systems or of alternative public domain networking solutions.

This has also become the subject matter for an expanding array of specialized industry standards consortia. These groups are often unconnected to the 'officially recognized' national or international standards institutions, and, although they may elect to seek a broad base of inputs, private consortia are under no compulsion to operate under any measure of public scrutiny.

The OSI experience would seem to contradict Verman's assertion that standards achieved by consensus in a publicly recognized institution have coercive power to ensure implementation. This observation, however, only holds if we define 'implementation' in very narrow terms. It can, with some assurance, be argued that OSI has been instrumental in establishing the concept of the open, multi-vendor computer networking environment among suppliers and users.

That OSI continues to be viable in any sense indicates that there remains a perceived need in both the public and private sectors for a central point of reference, even though there may be additional value in participation in fora that are more immediately responsive to temporal standardization concerns. Indeed, there are now signs that even historically ardent public sector proponents of OSI are contemplating a more pragmatic 'hybrid' approach to interconnection and interoperability (Carver, 1992).

It could be that initiatives like the ISO/IEC proposal for *quasi* standards would coordinate or even supplant the consortia phenomenon. Alternatively, however, it could also be that such proposals would become bogged down in their own complexity, and, in the process, act to diminish the stature of standards development organizations as independent bodies seeking consensus from among the widest possible community of interests. The question concerns the extent to which the latter alternative actually matters.

THE CONSENSUS IDEAL AND PUBLIC POLICY FOR STANDARDS

Many of the most significant recent public policy initiatives with respect to standards have come from the European Community (EC). These are partly the result of a concerted attempt to remove technical trade barriers between EC member states and to establish a 'single European market', but it is also clear that EC standardization policy is connected to various industrial policy objectives including the coordination of the R&D effort in the member states (European Community, 1985; Barry, 1990).

While generally insisting upon the non-mandatory nature of standards as developed in European national and/or regional standards institutions, several policy statements have given several significant indications of adjustments in EC thinking about the place of consensus and voluntarism in developing these standards.

Beginning in the early 1980s, policies were adopted establishing a mechanism whereby the Commission of the EC could 'prime' the voluntary/consensus process to some degree by issuing funded 'standards mandates' to selected regional standards bodies (European Community, 1983). In the 1987 'Green Paper on Telecommunications' the development of common standards for the evolution of a European telecommunication infrastructure was firmly established as a policy goal. The Green Paper recommended the founding of the European Telecommunication Standards Institute (ETSI) to coordinate standardization in support of a harmonized European technical infrastructure (CEC, 1987).

Although the voluntary/consensus principle became the backbone of the ETSI methodology, it is of interest to note that the authors of the Green Paper considered such methodologies to be unsuited to ETSI's purpose owing to the perceived inadequacy of the part-time deployment of voluntary resources in the light of pressures to accelerate the harmonization of the European infrastructure (CEC, 1987).

ETSI removed European telecommunication standardization from the domain of confidential agreements struck between monopoly network operators and their preferred suppliers. It has been noted, however, that the process of expanding the range of interests represented in ETSI also potentially made the possibility of coming to consensus that much more difficult (Besen, 1990). To compensate for some of the perceived inadequacies in the voluntary/consensus method, ETSI instituted various management and monitoring innovations, including the use of funded 'project teams' to work on specific technical problems.

The major indication of the somewhat ambiguous perception of the role of the consensus methodology by policy-makers in the Commission came in the 1990 'Green Paper on Standardization' (CEC, 1990). Responding to what it considered to be an unacceptably slow response by EC-recognized regional standards bodies to 'single market' standardization requirements, the Green Paper recommended specific structural changes aimed at increased 'efficiency' in the resourcing and production of standards. 'At a time when important decisions at the political level are taken on the basis of majority vote, there needs to be a shift away from an unqualified commitment to consensus in European standardization, although the Commission accepts that the use of standards is related to the degree of consensus reached in their elaboration' (CEC, 1990: p. 23).

If used as a basis for policy, however, this statement has some troublesome inconsistencies. First, it cannot be assumed that the political decision-making process is an identical phenomenon to the negotiation of technical agreements even though the two activities may be correlated. Second, it could be argued that consensus either exists or it does not – that questions of 'degree' serve only to negate the principle!

The Green Paper proposes a variety of 'efficiency' mechanisms which are almost solely related to increasing the speed of the standardization process. These include the establishment of sector-specific 'Associated Standards Bodies' and the resourcing of 'drafting secretariats' and 'project teams'. They also include imposition of 'majority voting' procedures for draft standards, shorter public inquiry periods and the use of information technology to reduce paperwork.

The Green Paper presents the 'managed standards-making' approach in ETSI as the most appropriate model upon which to build. The problem is that this approach has actually yielded very mixed results in ETSI. A number of major ETSI standards programmes have delivered their products within quite acceptable time-scales (even ahead of schedule in some cases) but have still encountered significant implementation delays. Either the promulgated standards do not attract wide industry confidence or they are otherwise difficult to implement owing to residual technical problems, regulatory impediments or certification problems. Committee participants have sometimes encountered significant problems distinguishing between technical and political agenda in ETSI (Hawkins, 1993b). Project teams have sometimes proven difficult to form, and often there are questions in the minds of ETSI members as to which interests the project team members actually represent.

The ambiguity regarding consensus is concentrated in the assumption that 'efficiency' in standards-making is equivalent to the speed and volume of document production. In terms of the political agenda of the 'single market' initiative, these criteria might well give the impression that the progress of this agenda is being quantitatively monitored. The net effect, however, is often to engender the general perception that European standards represent more of a political consensus than a technical and commercial one.

Historically, the institutional standardization structure in the US has been much more fragmented than its European counterpart. A recent US government study was critical of this situation, warning that failure to impose some measure of central coordination on standards would result in unfavourable conditions for international trade in US-made goods (US Congress, 1992). In contrast to the EC Green Paper, however, it is noteworthy that the US study squarely identified the central problem that no objective set of indicators currently exists which could be accepted commonly as a basis for the evaluation of the standardization process in terms either of identifying or achieving objectives (US Congress, 1992).

The difficulty with the definitions of standardization 'effectiveness' and 'efficiency' in the EC standardization Green Paper is that they assume that the benefits of a European 'single market' free of technical impediments to trade are sufficient rationally to commit European industry to support massive standardization initiatives (CEC, 1990). That European industry has not responded is perhaps not so much an indication that the institutional structure is inadequate, but rather that industry priorities with respect to standardization may not be sufficiently congruent with those of the Commission of the EC.

CONSENSUS AND EFFICIENCY IN CONTEMPORARY STANDARDS-MAKING

It is not the purpose of this Chapter to argue against making adjustments to the institutional processes of standardization. The proposition is rather that the basis for determining which adjustments are appropriate should reflect evolutionary trends in the purpose of these institutions. It is not simply a matter of determining under which conditions voluntary consensus-building methodologies might be considered 'efficient' in terms of eliciting outputs from standards organizations, but rather of determining in the first instance what range of functions participants expect these organizations to encompass.

Analysis from several quarters indicates that in the long term the documentary products of standards committees may be of less overall importance than the institutional position of the standards organizations themselves with respect to the development, exploitation and management of technology. Reddy (1987) has suggested, for example, that it is analytically problematic to consider standardization in terms of discrete technologies. Standards should be seen not simply as variables in the buyer–seller equation affecting individual products, but as 'coordinating mechanisms' in evolving technological systems. Standards institutions facilitate 'interorganizational, interindustry interdependencies' and 'informal cooperative action' between firms which is applied primarily at the generic product level (Reddy, 1987: pp. 51–8).

Stressing the function of committee standards in clarifying commercial relationships with respect to technological development, Cargill (1989) suggests that the goal of a standards committee should not be simply to create standards, but also to provide a forum for exchanges of technical information which are often necessary before industry-wide discipline over technical development can be imposed. Effective use of the process is dependent upon this understanding. In a similar vein, Schmidt (1992) sees standards committees as living and evolving mechanisms for the ongoing coordination of individual technologies into complex technical systems, thus injecting a measure of stability into an environment in which no single actor has overall control.

These observations are redolent of Partridge's sense of consensus as related to the support of 'systemic' relationships in societies. Standards committees can play important roles in the brokerage of positions and interests between public and private sector actors. The aims and achievements of committee standardization processes should be evaluated therefore in terms of their ongoing negotiation functions as well as in terms of the production and implementation of their documentation.

This diplomacy-like function applies most clearly to those established national and international standards organizations having an existing posture of public responsibility. It is reasonable, however, to suppose that it could also apply to *ad hoc* committees and consortia. The writer would argue that factors of compromise and concession come most actively into play in determining the technical level at which consensus might be possible, and the scope of the subject matter to which this consensus might apply. To use Zartman's terminology, this involves the separation of 'formula' from 'implementing detail'. The difficulty for standards institutions lies in the fact that increasing complexity may make it unlikely that the formulae and the details can be identified in the same forum. A fully articulated standardization process could encompass the dynamic selection of institutions by various groups of actors depending upon whether the need is for formulae or for detail.

Such a proposition suggests a re-evaluation of how to research the ubiquitous questions: 'What is a timely standard'? and 'How many standards are required?' Theorists have drawn attention to the many discrepancies which exist in perceptions of time and their relationship to rationality in social decision-making (Hayden, 1987). Recent experience suggests that the assumption of congruent motives on the part of standards committee participants is misplaced. Standards-making is now not so much a process of coming to consensus about how existing or future technical practices should be codified, it is more an organic process of determining whether the basis for consensus exists at all, or if it can be built, and with respect to which elements.

Time-scales can be expected to vary according to the expectations and agenda of the participants, and according to the chosen institutional settings. At one extreme, formula definition requires discussion of non-technical as well as technical issues and likely needs to be accomplished in a broadly based forum. At the opposite extreme, individual technical details can be addressed by highly specialized fora in which the scope for dispute is relatively small.

Relationships among traditional standards bodies and consortia must be viewed against the possibility that there are a growing number of requirements in the standardization process and that more than one institutional structure may be necessary to fulfil them all. The challenge for public policy is to see that the public interest does not become lost in this diversity. It is not just a matter of supporting and monitoring standards initiatives, but also of more clearly defining the terms under which standards bodies in the contemporary *milieu* will be considered to be acting in the public interest.

The strength of the committee process for standardization as it has evolved in this century is embodied in the principles of consensus and voluntarism. At the most basic level, these principles have been necessary preconditions for the provision by industry of the required technical and human resources. The failings of the system have revolved around achieving an effective balance of representation in what remains essentially a supplier-led institution. Nevertheless, where the premise that a standard represents a 'voluntary consensus' is brought into question, the legitimacy of the institution that produced the standard is also undermined.

The dilemma for policy makers is really one of instrument selection. Governments can opt for judication where standardization problems present impediments to their economic and social objectives. Alternatively, they can delegate a measure of authority in these matters to non-governmental organizations. The problems arise with the application of pressure for structural and procedural changes in voluntary standards organizations, which may tip this activity towards less equitable and responsive modes of decision-making.

As with diplomacy in the political realm, there is a strong case that voluntary/consensus standards negotiations should be allowed to find their own pace and proceed accordingly, provided that a reasonable amount of equity in the process can be guaranteed. A challenge for policy-oriented research is to construct an analytical framework that defines the institution in terms of the complex and diverse expectations of the participants in its activities.

PART 3

The political economy of standards, innovation
and competitiveness

15. The roles of standards as technology infrastructure

Gregory Tassey

In technology-based industries, the roles of standards are ubiquitous. A majority of these standards are based on sophisticated technologies and therefore require substantial research to develop. Moreover, because they are jointly used by competing firms, standards qualify as one type of economic infrastructure. The strong public good character of infrastructure means that government frequently plays a role in its provision. This is the case with respect to providing the technical basis for standards, even those promulgated by industry on a voluntary basis.

Industry standards affect product and non-product elements of an industry's technology in a number of ways. Firstly, they perform four direct functions: (a) to reduce the variety of a generic product to enable economies of scale to be achieved; (b) to provide accepted methods of producing information, such as measurement data for process control or test data for consummating market transactions; (c) to prescribe acceptable levels of performance, including minimum levels of quality, safety, and so on; and (d) to enable compatibility or interoperability among components of a system (Tassey, 1992: chapter 6).

The technical bases for all but variety reduction have a large if not complete public good content. These other three categories are directed at non-product technology elements (interface protocols, measurement and test methods, evaluated science and engineering data bases, and so on), which are on average more difficult for individual firms to make proprietary. Government, therefore, takes a major role in the provision of many of the infratechnologies that are the basis for standards (Tassey, 1992: chapter 3). In the US, the majority of these standards are promulgated by industry through voluntary standards committees. Government often has a representative on these committees, usually from the National Institute of Standards and Technology (NIST), to provide technical input and facilitate the standards-setting process, but the final decision lies with industry.

The variety reduction function, on the other hand, is often achieved by one firm gaining control of the underlying technology and thereby forcing other competing manufacturers to adopt it (often through licensing agreements). The product element then becomes a *de facto* standard.

In a technology-based economy, both product element standards and standards based on public good technologies (infratechnologies) are taking on increased importance due to the complexity of new products and the fact that these products are frequently embedded in systems. The productivity of these systems, whether they be the basis for production processes (factory automation) or the provision of services (communication or information processing), is dependent not only on the effectiveness of individual system components, but also on the interfaces between these components. Standards play important roles in both categories of productivity determinants.

Moreover, the evolutionary pattern by which the final structure of standards in a production chain is attained has a significant impact on the fortunes of individual firms, industries and hence world-wide competitive positions. The pattern of standardization at one stage in a production chain (such as components) affects strategic decisions concerning earlier stages in the production process (such as materials) and later stages (such as equipment, systems and services).

COMPETITIVE DYNAMICS

The magnitude of global competition is making control by one firm of infrastructure elements at several stages in a production chain difficult, if not impossible. This is particularly the case for standardization at the interfaces between product elements and products that comprise a system. Even so, many firms are not waiting for the market's competitive dynamics to eventually create an environment in which consensus standardization can finally take place.

For example, 'open' technology-based product systems have the substantial efficiency advantage of allowing an entrepreneur to market an improved product or service without having to create a complete vertically or horizontally integrated system. Conversely, the user can custom-design a system by choosing components from various manufacturers, with the result that system performance can be optimized with minimum cost.

The technological innovator that chooses to promote its technology as 'open' sacrifices some control over the future direction of the technology and some opportunity to integrate horizontally and vertically (that is, to sell 'turnkey' systems, with monopoly control over replacement sales). In return, however, the innovating firm becomes the initial technological leader of a much larger market, one with a greater probability of lasting.

On the other hand, the existence of standards, while achieving the several important categories of economic benefits discussed here, can have a retarding effect on innovation. For example, if Microsoft, Apple and IBM were freed from maintaining compatibility with existing hardware and software, new generations of system software would evolve at a much faster pace. This phenomenon is one reason why smaller firms appear more innovative, especially with respect to more radical innovations. However, success as an innovator manifests itself in a large installed base. Once a large customer base is attained, it must be served and this requires an evolutionary rather than a revolutionary approach to change, including adherence to existing standards.

Thus, technology-based competition involves a constant battle among firms for significant shares of components markets, who then want their protocols to be at least *de facto* standards in order to stabilize, if not increase, those market shares. While such standards may be a deterrent to radical innovation, lack of other types of standards affecting Research and Development (R&D) efficiency, quality control, transaction costs and the interfaces between components of a system can be a serious deterrent to long-term growth and achievement of a competitive position.

In general, standards are essential to the efficient evolution of most technologies. In fact, to the extent a technology has a systems structure, a large number of standards must be in place simultaneously before the relevant markets can grow. The computer industry is a good example of the benefits of standards, as well as the problems that arise as the standards evolve. Similar statements can be made for communication networks, factory automation and so forth.

In the early period of a technology's evolution, innovative firms attempt to dominate markets with proprietary versions of the generic technology. As markets grow, realization of economies of scale requires that certain product element designs gain control as *de facto* standards. Interoperability requirements cause users to increase pressure for 'open' systems. The resulting competitive dynamic initially forces partial standardization in the form of several industry segments with their own standards.

Having a few standards, as opposed to a larger number of proprietary product elements, interfaces, and so on, only partially mitigates the problem for buyers. In fact, the failure of expectations for standardization to be realized leads to confusion and reduced demand. Eventually, because no one firm or coalition of firms typically ends up dominating the market, a single set of standards finally evolves.

The costs of letting this process take place entirely through market forces can be high, especially when the domestic industry finds that foreign competition has coalesced behind a single set of standards that is not optimal for the domestic industry. Slowness in agreeing upon certain types of standards can allow competition to more rapidly penetrate the global marketplace through the efficiency gains that standards can create.

NON-PRODUCT STANDARDS

US industry currently has approximately 40,000 non-product standards in use. (Government sets an approximately equal number of standards, a majority of which are military procurement specifications.) The technical bases for these standards have a large public good content. Virtually all of these underlying technologies (infratechnologies), and therefore the resulting industry standards, are derived from basic standards – the fundamental constants of nature. Basic standards represent the most accurate current statement of the fundamental laws of physics and have such diverse applications that they qualify as pure public goods and hence are provided entirely by government. Figure 15.1 illustrates an example of the hierarchy through which basic standards are utilized to develop infratechnologies upon which industry standards are based.

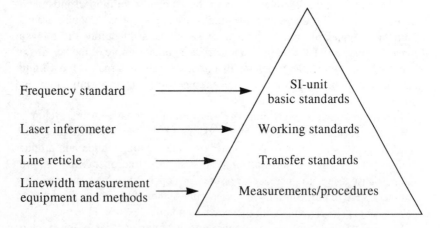

Figure 15.1   Path to an industry standard

The production of semiconductor components is a highly demanding process. The densities of today's circuits are such that each conducting path ('line') on a chip is a small fraction of the width of a human hair. These widths must be consistent with respect to design specifications to avoid thermal, electrical and other problems. The semiconductor producer therefore needs to be able to measure the widths of circuits that make up chip designs.

Particularly important are linewidths on the 'masks' which are used to inscribe the multiple layers of circuit patterns on the chip itself. Their quality greatly affects performance of the chips produced and hence a semiconductor manufacturer's production yield.

The linewidth measurement equipment must be calibrated against a physical standard, which has a pattern of lines with thicknesses and spacings determined to a specified level of accuracy. This determination is done by an authoritative source, such as NIST in the US. The physical or 'transfer' standard used by industry must be easily transportable (a 'reticle' in the above example) in order to ensure widespread and accurate transfer of the infratechnology.

The information transferred by a physical standard is itself determined or certified by a so-called 'working' standard, which is laboratory-based and therefore not readily transferable. In this example, the working standard is a laser interferometer which measures and certifies the physical dimensions of the line reticle prior to transfer to industry. Finally, the laser interferometer is itself dependent for calibration on a 'basic' standard for length.

THE PUBLIC ROLE

Because of the high public good content of most non-proprietary standards, government funds significant portions of the underlying infratechnologies. Industry also contributes to these infratechnologies, especially at early stages in a technology's life cycle when customers are demanding measurement or test methods before a standards-setting process can be carried out, or before government research can provide the underlying infratechnologies. Increasingly, sophisticated customers demand assurance of performance or systems compatibility before making even limited purchases. Without a standard method for providing this assurance, individual firms develop proprietary procedures, which often vary enormously.

Multiple measurement and test methods based on the same general infratechnology usually increase transaction costs substantially as companies expend skilled labour and other resources attempting to resolve disputes over performance verification. These transaction costs add to the overall cost (and hence price) of a new technology and thus slow its diffusion into the marketplace.

Some infratechnologies are developed entirely within government laboratories and others are simply selected from existing alternatives in the private sector and used as the basis for the standard. However, with product life cycles becoming ever shorter in response to increasing global competition, the timing of the promulgation of standards becomes more important. Thus, other mechanisms are being increasingly used that respond to the increased competitive pressures and also more accurately respond to the '*quasi* public good' nature of many infratechnologies.

One fairly recent response to the timing and 'public good' concerns has been increased cooperation between government and industry to produce the infratechnology earlier in the technology life cycle, or at least in the time frame dictated by market strategies. As of March 1994, NIST had 14 research consortia with various industry groups pursuing such infratechnologies as process control databases for advanced materials such as ceramics and polymers, quality management techniques for automated analytical laboratory systems and test methods for microwave integrated circuits. Ongoing research is examining the strategic motivations of firms for participating in these research consortia (Cordes and Tassey, 1993).

ECONOMIC IMPACTS

The cost of not developing infratechnologies and promulgating standards based on them is high. Without sufficient measurement accuracy, R&D cannot advance scientific and technical knowledge and without standardized versions of the required measurement methods research results cannot be effectively conveyed or transferred. Without standards for measurement of process parameters in a production system, automation cannot be advanced and quality cannot be assured for the purchaser. Similarly, post-production testing to verify a producer's performance claims cannot succeed without standardized acceptance test methods.

A number of studies of the economic impact of infratechnologies and the standards in which they are embodied have been undertaken at NIST. This impact is estimated using an internal rate of return calculation, identical to what economists call the Social Rate of Return (SRR).

Table 15.1   Social rates of return from government investment in infratechnologies

Technology area	Social Rate of Return (%)
Semiconductor – three sub-areas	63–181
Optical fibre	423
Semiconductor – electromigration	117
Electromagnetic interference	266

Table 15.1 summarizes quantitative estimates from these economic studies. The methodology is consistent with the few other impact studies of either government or industrial R&D. The methodology used by NIST is also consistent with that in earlier studies of the SRR from basic research. The SRRs from NIST research are high compared with those found in studies of cross-sections of private investments in technology and other public technology investments. These results are not surprising, given that NIST targets its research at specific industry infrastructure problems and builds a technology transfer strategy into each project.[1]

Assuming the conventional negatively sloped marginal efficiency of capital curve, the implication of these high SRRs for investment in infratechnologies is that this category of investment in technology infrastructure is too low relative to private sector investment in technology.[2] Expansion into related areas, presumably with somewhat lower SRRs, would reduce the average SRR for the project area as a whole until the rate of return approached the SRRs for other types of investment – at least for other types of technology investment (Link and Tassey, 1993).

PRODUCT ELEMENT STANDARDS

As opposed to the above class of non-product industry standards derived from public good infratechnologies, product element standards are frequently based on proprietary technologies. The resulting standards are typically *de facto* rather than promulgated by industry consensus and usually result from intense competition.

One of the more visible examples of the competitive effects of *de facto* standardization is the 'architecture' of personal computers. 'Architecture' is the set of standardized product elements and the rules or protocols for their interaction with other product elements. Such elements of computer architecture as the operating system, the 'bus', the graphical user interface and the applications programming interface, have been the focus of intense competition, although the efficacy of competition to supply the product elements comprising an architecture is still a matter for debate (Morris and Ferguson, 1993; Aberthal *et al.*, 1993).

For example, Apple made a brilliant move when it forced third-party software developers to adopt a standard 'graphical user interface', so that all programs running on Apple computers presented the user with the same screen format and command structure. However, Apple kept its hardware operating system proprietary, which meant that Apple would prosper or fail entirely on its own ability to maintain sufficient market share.

In contrast, Sun Microsystems 'opened' the microprocessor architecture for its workstations in order to obtain 'help' in maintaining market share. It was willing to share its markets in return for the increased probability that its architecture would become a standard.

INDUSTRY STRATEGIES

In the early years of the computer industry, standardization focused on physical interfaces and, later, on the functional interfaces between operating systems and applications programs. However, as the number of computer users grew and, especially as the large computer users accumulated many different types of computers, the demand increased rapidly for standards to support the broader concept of 'interoperability' – enabling computers made by different vendors to communicate.

Users could invest in integrating different vendors' products themselves. This strategy was expensive and time-consuming and it led to rigid configurations that became obsolete – either technologically or with respect to new corporate market strategies. Frequently, companies could not bear to scrap their investment and simply accepted the inefficiency of an out-of-date information system.

In concept, the 'open systems' computing paradigm allows users to combine different vendors' products across several platforms without investing in custom integration. In the ultimate application of this concept, heretofore largely separate worlds of R&D, manufacturing and management can 'interoperate' as one system with 'seamless' connections between its many components.

In practice, achieving truly open computer systems has been extremely difficult. Compatibility has been partial at best. One of the major reasons for this difficulty is complexity of systems technology. Each one of a number of layers in the hierarchical structure of a typical information system must have its own architecture and these architectures must be integrated into a network. Today, systems technologies not only have complex hardware and software platforms, but rules and protocols must be in place to introduce, process, store, move and retrieve information. The systems architectures are typically controlled by a handful of firms. Competitors are constantly vying for these monopoly or *quasi* monopoly positions. Along with this competition for the technology content of product-element standards, the evolution of non-product protocols and standards is taking place.

Vendors often claim to conform to 'open systems' requirements, but in reality they conform only to a portion of the set of standards to which compliance is necessary. Moreover, even when true open systems compatibility is offered, users often find that 'mobility' from their current proprietary system to the open system does not exist, even when the same vendor provides both. Transition products are not made available, leaving the user with a severe obsolescence problem.[3]

From both the corporate strategy and public policy points of view, standardization is not an all-or-nothing proposition. In a complicated systems technology such as distributed data processing, telecommunication or factory automation, standardization typically proceeds in an evolutionary manner, with the pattern of evolution being determined by: (a) the pace of technological change in each component category; (b) the disembodied technology development, which determines the overall system architecture and organization; and (c) the market structure which includes the incentives and abilities to force the standardization process.

The degree, as well as the timing, of standardization is important for the effective market penetration of a technology. In the case of numerically controlled machine tools, for example, total standardization of data formats would have severely compromised the range of performance attributes desired by different users in the machine tools they purchased. Thus, only a 'degree' of standardization has been optimal, at least up to this point in the evolution of the technology.

As technology-based systems become increasingly important and the 'windows of opportunity' for making successful investments in the associated markets continue to shrink, the relevant standards will have to be 'managed'. If a standard is fixed, even if it is competitively neutral (an interface standard, for example rather than a proprietary product element standard), it will eventually act to stifle the introduction of new technology into the system. Alternatively, if the standard is updated through successive revisions, then version consistency (upward mobility of existing system components) can become a problem.

For example, many observers view the future of data processing as being accomplished by large distributed systems, which are only possible if standard interfaces are provided on all the communication paths in the network. With new technology continuously being introduced into ever expanding networks, the pressure on the standards infrastructure to evolve is substantial. Standards must be continuously updated, not only because of changes in distributed processing technology *per se*, but also for innovations to address toleration of faults, rapid (real-time) response to multimedia inputs and outputs (voice, image, text, and so on), human friendly interfaces and adaptation to a constantly changing network structure as nodes are added, removed or replaced by different systems.[4]

CONCLUSION

Over the past decade, the infrastructure roles of standards have increased in importance because: (a) the demand for quality and performance in technologically complex products and systems requires a range of standards based on sophisticated infratechnologies; (b) the systems nature of critically important technologies means that competition is greatly affected by the degree of standardization at the interfaces between components; and (c) the shortening of the average technology life cycle has on average increased the pressure on the standards-setting process with respect to the optimal timing.

The management of these infrastructure roles is driven by the fact that the interaction of the standardization process with industry behaviour and structure is a recursive relationship. Therefore, opportunities exist over much of the typical technology life cycle to adjust the standardization process in ways that an industry's competitive position is enhanced.

From an economic growth efficiency point of view, not only will competitive positions suffer if the appropriate infrastructure, including standards, is not in place at the right times in a technology life cycle, but leading-edge users may invest in their own proprietary infrastructure, creating an installed base that inhibits a more uniform and efficient infrastructure for the entire industry.

The central strategic problem of managing the timing and content of standards is difficult partly because of the many types of standards needed in today's typical technology-based industry and partly because standards, as one type of technology infrastructure, are derived from different sources than the industry's core technology and often conflict with individual corporate strategies.

Diffuse decision-making through multi-enterprise market structures drives innovation and thus provides a rich initial technology base, largely without standardization, although certain categories of standards are essential to conduct the R&D that creates and diffuses the technology. However, the lack of standardization can greatly inhibit technological progress through the middle and later phases of the technology life cycle. A concentrated market structure coupled with aggressive government intervention could, superficially at least, solve the problem of insufficient standardization, but creativity would suffer. The answer seems to lie in applying an understanding of how technologies evolve over time and supplying industry consensus standards that fit each phase in the evolution of a technology.

NOTES

1. The success of these infratechnology research programmes depends in part on the availability of a measurement science base. Thus, basic research conducted in industry and at universities, NIST and other government laboratories, is an essential precursor to achieving high rates of return in all technology research. Mansfield (1991) estimated the SRR to 'academic research' that contributed directly to the commercialization of products and processes to be 28 per cent (this later number is smaller than estimates produced in other studies because of the relatively narrow scope of impact being measured and the relatively long lags to which a discount rate is applied). The median estimated rate of return from studies of private-sector innovations is in the 50–70 per cent range.

2. The marginal efficiency of capital curve represents the typical pattern of resource allocation by which limited resources are first allocated to the projects with the highest expected rate of return, then to projects with somewhat lower expected rates of return, and so on, until funds are allocated to those projects with expected rates of return just equal to the 'hurdle' rate, that is, the rate obtainable in other areas of investment (also referred to as the 'opportunity cost' of not investing these resources elsewhere). Note that the hurdle rate includes a risk premium, which will be higher for the typical R&D investment compared with the typical investment in plant and equipment. Thus, the hurdle rate will be higher on this basis alone for R&D projects. However, the *ex post* SRRs to R&D are sufficiently high that the expected SRRs on many R&D projects exceed the hurdle rate, even with the risk premium, and therefore both industry and government invest substantial sums in R&D. For the past six years, Japanese industry has invested *more* in R&D than it has in plant and equipment (Kodama, 1991). This latter trend says a lot about relative expected rates of return in Japan.

3. As an example of the complexity of the open systems concept, the Institute of Electrical and Electronics Engineers (IEEE) definition includes Portable Operating System Interface for computer environments (POSIX), the XPG3 portability standard, the International Organization for Standards (ISO) communication standards, and American National Standards Institute (ANSI) compliance standards.

4. Japanese industries, motivated by Japan's world-wide lead in consumer electronics, have been developing an operating system called TRON, which is designed to integrate extremely diverse systems of intelligent objects that are networked together (Kahaner, 1991).

172-91

16. The political economy of data communication standards

W. Edward Steinmueller

INTRODUCTION

Fibre optics and related technologies are the basis for recent proposals to construct 'data highways', high-capacity data networks that have the potential to radically transform data telecommunication services and create an entirely new structure of networks of information services for businesses and consumers. As in any radical technical change, the chances of arriving immediately at a socially optimal market organization or technological configuration, are essentially nil. Moreover, any 'data highway' plan will involve extensive regulatory deliberation and intervention by a regulatory structure that has evolved over the past century to solve problems quite different from those that are likely to be encountered in building new data services and information networks. The financial stakes of implementing data highways are large because financing the high fixed costs of deploying fibre optics requires a revenue flow that can only be achieved through a rapid utilization of network capacity. Delays in the utilization of capacity have the potential to create large operating losses or impose high rates on customers. Despite these apparent problems, there is widespread optimism about the social and economic value of enormously increasing data communication capability. Whether this optimism is warranted, or is similar to the optimism that led to the promotion of nuclear power as the source of electricity that would be 'too cheap to meter', will depend upon how successfully the market is organized to generate demand for new capacity and services, the central concern of this Chapter.

Given the strong public role in telecommunications and the purported contributions of data communication to 'international competitiveness', it appears likely that virtually all industrialized countries will undertake large-scale investments in data highway projects during this century. Since we live in an era where 'fiscal stringency' is a nearly universal context for considering new public investment initiatives, prospective investments in data highways are likely to be based on confident projections that future service revenues will justify investment costs. This Chapter advances two principal themes. Firstly, achieving capacity utilization and therefore adequate revenues for a 'data highway' is, in fact, an uncertain prospect because of the shortcomings of centralized decision-making in regulation and service offerings. Secondly, the prospects for capacity utilization may be improved by new approaches to the technical standards-setting process that support coordination among firms while decentralizing decisions about specialization and interfirm division of labour. The latter thesis, the need for a new approach to standards, is based on two premises: (a) that the pace of technical change, complexity of the requisite complementary technologies and difficulty of organizational reforms are likely to overwhelm existing telecommunication organizations; and (b) that high levels of capacity utilization will require the sort of risk-taking and variety that characterize competitive high technology markets. These themes stand in direct opposition to the most likely course of development, one involving the use of centralized decision-making to dictate the standards and offerings of data communication and information services using fibre optic telecommunication networks.

This Chapter's concern for the inadequacy of existing regulatory and market approaches for data telecommunication is based on the recent history of regulatory decision-making in computer and telecommunication markets during which these markets have become more interrelated. The move toward deregulation has substituted the expectation that coordination through market competition would replace centralized coordination in the provision of telecommunication services. This has meant that a primary focus of standards-setting efforts has been the concern for 'open networks' and the amelioration of interconnection costs rather than any fundamental rethinking of what will be the nature of coordination, for example, issues of interoperability or 'end-to-end' solutions for users, under market competition. Major mechanisms of market coordination under competition are the result of interactions among: (a) the interests of large customers that are increasingly able to 'bypass' the public network; (b) independent equipment producers who seek to maximize their profits in the supply of equipment; and (c) telecommunication service providers whose service offerings are determined by regulatory processes.

Standards-setting is a principal method for coordinating the supply and demand of equipment suppliers, service providers and customers. The standards-setting community (dominated by corporate technical representatives from each of these groups) has played a growing role in defining service offerings. Although this technical community has successfully introduced a number of important new standards, regulators have found it difficult to keep up with technological opportunities in their decisions about service offerings and tariffs and thus have sometimes placed telecommunication service providers at a competitive disadvantage. Among the reasons for regulatory delay and indecision are the facts that greater heterogeneity of data communication services provides a bewildering range of regulatory options and that such services offer much more complex pricing and costing problems than voice lines. Public regulation of private monopoly or government enterprise decision-making about data telecommunication services is therefore likely to be more costly, subject to greater delay and less effective in choosing which services to offer for data telecommunication than ordinary voice telecommunication (Steinmueller, 1994).

The goals of a decentralized approach to standards-setting in data telecommunication services are to accelerate decision-making, facilitate the division of economic activities among firms, and provide the advantages of market competition in developing new data telecommunication services. A key feature of the decentralized proposal is a novel approach to the standards-setting process that strives simultaneously to preserve the autonomy of competitive service offerings and coordinate the service offerings of competing firms. To evaluate the feasibility and barriers to a decentralized approach it is useful to briefly review existing theories about the existence of the firm, vertical integration and the division of economic activities among firms.

FIRM AND MARKET ORGANIZATION

In economics, the theory of market organization, that is, the explanation of who produces what and for whom, is sufficiently problematic that most economists choose to take a current market organization as given and attempt to explain changes in it such as vertical integration or horizontal combination. While this approach is useful in explaining interactions between firms, it only indirectly addresses questions about the boundary of the firm, namely why some economic activities are conducted internally and others are relegated to suppliers or customers. The first step toward a more comprehensive theory is to determine a rigorous basis for the existence of firms, that is, why should firms exist in theory as they do in reality? As Coase (1937) noted, if factor allocations and the production levels are determined by market price, why is it necessary to have an organization – the firm – that performs additional coordination activities? Indeed, it would appear, at first glance, that additional coordination activities within the firm would place it at a competitive disadvantage if such internal coordination activities utilized resources that were redundant to the coordination offered by market mechanisms. Examining the economic basis for the location of these additional coordination activities *within* the firm offers clues about coordination methods that will be useful in organizing economic activities *among* firms.

Coase's own explanation of the existence of the firm begins with the observation that in establishing non-market coordination through the entrepreneurial function the firm supersedes the price mechanism (Coase, 1937). This suppression proves to be desirable because there are costs involved in using the price system for coordination purposes. In short, Coase argues that the firm exists because entrepreneurs and managers accomplish allocative decisions more efficiently than the market. This greater efficiency arises principally from the fact that market exchanges are subject to transaction costs, the costs of establishing contracts that determine value and provide redress for non-performance. Coase also applies this idea to the problem of firm growth by noting that there are diminishing returns to the entrepreneurial function as firms become larger. These diseconomies raise the costs of internal transactions relative to market transactions until the further expansion of the internal organization of economic activities is uneconomic (Young, 1928; Penrose, 1959). From this argument, a serviceable theory of vertical and horizontal integration may be constructed (Williamson, 1971). A primary deficiency in deriving policy implications from this theory is the difficulty of observing or evaluating transactions costs. For example, it is difficult to know where, or to what degree, inter-firm coordination would be facilitated by public subsidy because we do not know how close a given subsidy will bring the firm to the margin at which it conducts more transactions with coordinating objectives. Such a subsidy may be employed for some other purpose or, if tied to inter-firm activity, may either be ignored or serve as a means of increasing joint rents rather than coordination.

For technologically sophisticated products such as data telecommunication services, the definition of product creates further complications for the comparison of intra- and inter-firm division of economic activities. The absence of contingent markets for technological alternatives requires that firms learn much more about their suppliers' capabilities and their customers' requirements than is available in the sort of market economy visualized by neoclassical theory. Nonetheless, the additional information requirements for participating in these markets can be accommodated within the transactions cost theory approach. In doing this, however, we introduce another set of unobservables that dictate the actual performance of firms.

It is essential to recognize that Coase's argument was aimed at establishing a rigorous analytical basis for the *existence* of the firm and, as a by-product, offered a theory of the limit to the expansion of the firm. There are three features of Coase's original formulation that require closer scrutiny for our purpose of constructing a theory of the division of economic activities among firms engaged in producing data communication services and constructing information networks. The first of these involves the impact of exogenous changes in the costs of coordination. Coase argued that improvements in communication technologies (he cited the telephone and telegraph) 'which tend to reduce the cost of organizing spatially will tend to increase the size of the firm', but in a footnote he modifies this conclusion with the more precise statement, 'if the telephone reduces the costs of using the price mechanism more than it reduces the costs of organizing, then it will have the effect of reducing the size of the firm' (Coase, 1937: p. 25, footnote 31). Thus, so long as we take changes in the costs of telecommunication technology as exogenous, the Coasian framework offers a theory of the division of activities among firms according to the relative costs of internal organization versus recourse to market mechanisms. For our purposes, exogeneity may be achieved not only through the cost reduction of data telecommunication services, but also through cost reductions from standards-setting activities.

The second feature is whether it is appropriate to regard improvements in coordination costs and managerial technique as exogenous to the operation of the firm. Perhaps these improvements are endogenous to the level and collection of activities undertaken by the firm rather than delivered by an unspecified exogenous agency. If so, then we may properly introduce an additional coordination mechanism, 'strategic planning', to account for firms' allocative decisions and growth (Teece, 1987). The ability of the firm to reduce production costs with experience and learning economies is now commonly regarded as a source of first mover advantage contributing to the growth of firms and providing a basis for strategic planning (Coase, 1972). Differences among firms arising from their relative capacities for learning from past experience are unlikely to be observable *ex ante* and the growth of firms is therefore likely to rely on the accumulation of favourable competitive outcomes in which the firm demonstrates its superior learning capabilities. Confining growth achieved in this manner to a single firm requires the additional assumption that what is learned cannot be readily transferred outside of the firm (Winter, 1987). Otherwise, the market will bid away individuals that can transfer such information and endogenous changes that would have been the source of efficiency advantages for a single firm are translated into generalized advance in the industry. This raises an important problem for standards-setting activities. If standards erode or displace the competitive advantages from learning and specialization they are likely to be resisted by potential adopters.

The possibility that endogenous improvements in coordination costs and managerial technique may underlie the heterogeneity of firm performance suggests a re-evaluation of the absence of asset specificity in the Coasian framework; this absence is a third feature of the framework. Coase (1937) specifically rejects asset specificity as a causal mechanism for vertical integration. His point of departure is Klein, Crawford and Alchian (1978) who argue that vertical integration arises from asymmetries between supplier and customer when investment in specialized capacity is required. Under these conditions, when the customer has alternative sources of supply there is the possibility for opportunistic behaviour; supply costs may be forced downward by exploiting monopsonistic power that comes from holding the supplier's specialized assets as hostages in negotiations. In anticipation of this possibility, the supplier may elect not to create the specialization desired by the customer and the choice of internal production, namely vertical integration, will be pursued by the customer. When the customer has no recourse to other suppliers, the balance of control may favour the supplier and *quasi* rents may be generated at the expense of the customer due to the monopoly power of the supplier. In this case, the usual argument regarding successive monopoly applies, namely that the coincidence of gains in both social and private welfare supports vertical integration. Neither of these outcomes is pre-ordained (Coase, 1937). In both cases, as Coase notes, the superior outcome may be a long-term agreement based on joint profit maximization and a mutually agreeable division of any rents that derive from downstream demand. These same arguments apply if costs are endogenous. The problem is that many modern economies limit joint profit maximizing activity as an instance of anticompetitive behaviour and it may prove difficult to distinguish between anticompetitive and efficiency-based long-term agreements. The ability of firms to establish joint profit maximizing contracts is explicitly limited in the US by antitrust law. Procompetitive policies, such as the Robinson–Patman Act or 'cross-ownership' restrictions, may favour contracting over internal production. Given these limitations, standards-setting activities may be seen as a means of coordinating activities for either superior efficiency or enhanced market power.

In summary, the Coasian framework offers three specific hypotheses about the relative extent of vertical integration and contracting. Firstly, exogenous reductions in the costs of coordination will favour contracting over vertical integration. Secondly, endogenous improvement in coordination and managerial capabilities will favour vertical integration over contracting. Thirdly, 'competition policies' may eliminate the ordinary neutrality of asset specificity in the contest between contracting and vertical integration, either favouring or discouraging vertical integration depending on the specific content of the policy.

A fourth hypothesis is required to deal with the specific character of high-technology markets. This hypothesis is that users are likely to prefer a competitive market for products in which fundamental technical uncertainties are present. This preference is an implication of the fact that technologically sophisticated markets are characterized by fundamental uncertainties about the technological feasibility of future product offerings (including improvements in existing types of products). Final customers, or customers who purchase such goods as inputs, will prefer competition because such markets are characterized by greater variety and therefore a broader distribution of price and quality outcomes from which to choose. Since both types of customers are free not to purchase goods and services that they find too expensive or of insufficient quality, their welfare is improved by the choice offered by competitive market organization. Moreover, some of the losses from technical failure are borne by parties other than customers, a powerful motive for vertical disintegration of such activities. Ordinarily, the preferences of users for a particular market structure are irrelevant since their purchase decisions tend to produce whatever market structure best serves their interests. In the case of regulated markets or markets where purchasers have significant market power, however, this preference may be expressed through the political process of regulation or by negotiations between suppliers and the large customer.

Issues of political economy, the determination of economic outcomes through political process, enter into this analysis in all four of the hypothesized determinants of market structure. The first two, 'Coasian,' determinants of market structure are influenced by the supply of public goods that are complements to either exogenous reductions in coordination costs or endogenous improvements in coordination and managerial capabilities. For example, many technical compatibility standards are public goods that are exogenous and cost-reducing. Hence, the creation of technical compatibility standards may serve to increase the prevalence of contracting over internal supply. Similarly, public education, in managerial methods such as 'total quality control', provides a public good that will often be a complement to endogenous improvements and thereby favour vertical integration over external contracting. Public investments in promulgating technical compatibility standards and new management methods are therefore non-neutral with respect to industrial structure and are instances of the determination of economic outcomes by political means, that is, of political economy. The latter two hypotheses about the determination of economic outcomes are more directly political in character. Political decisions about competition policy, the third hypothesized influence on market structure, have an obvious and direct effect on market structure. Similarly, when regulation is used as the instrument for enforcing customers' (small or large) collective preference with regard to market structure, (the fourth hypothesized influence on market structure), the process is one of political economy.

This theoretical discussion sets the stage for considering the impact of standards in the organization of the two principal economic activities in data telecommunication, data communication service offerings and the construction of information networks. In both of these activities, inter-firm contracts and vertical integration are significant, that is, the division of economic activities among firms and extent of specialization are important. The next section provides an introduction to the economics and technology of these two activities and the reasons why I have identified them separately.

DATA COMMUNICATION AND INFORMATION NETWORKS

Data communication involves at least two termini, one of which is some type of computer, and a channel over which information, typically encoded as a digital stream of bits, is exchanged. At the most basic level, data communication involves the 'plumbing' through which information flows. Like plumbing, data communication requires control mechanisms for suspending and resuming flow, regulating the rate of flow, and so forth. There also may be differences in the size of data communication channels and the speeds of data flow, that require additional equipment. The main characteristic of economic relevance is the capacity of channels. A large share of the volume of current data communication flows over the voice channels used to deliver Plain Old Telephone Service (POTS). POTS channels are defined according to a technical reference standard that limits their capacity. The capacity of POTS channels is chosen as a trade-off between cost and quality. The current standard is the 3000 Hz, between 300 and 3300 Hz in most telecommunication systems, a frequency range suited to the human voice. The POTS technical reference standard has been used throughout the voice telecommunication network as a minimum performance specification for components. Although a given component may be capable of higher capacity than the reference standard, it is impractical to rely on higher capacity for reliable communication since each call connection may involve routing through different components somewhere along the channel. The theoretical limit for communication over a noiseless voice channel with no electronic noise to obscure a signal is 30,000 bits per second, a rate that for both economic and technical reasons is unattainable.

In addition to publicly switched voice channels, telephone 'common carrier companies' in the United States (PTOs in Europe) offer additional channel offerings such as leased lines. Some leased lines, such as those defined using the T-1 standard, offer higher data communication capacities than POTS lines. POTS retains the advantage that it may be employed to connect any two of over 500 million termini in the world voice telephone network. In addition, packet-switched data networks support data communication in the US and Europe. Europe has a large public-switched packet data network which, including terminals of all types, was estimated to have 500,000 termini in 1986 (Ungerer, 1990). Europe, the US and Japan have a variety of

Value Added Network Services (VANS) providers that offer switched connections (connections defined at the time of use) for their subscribers over leased or dial-up lines or, in Europe, over the packet network.

The market structure of data telecommunication channels is not a competitive outcome, but one which has been determined by regulatory mandate. Regulation has shaped the basic standards for POTS and enhanced leased line services (such as T-1 lines) as well as the detailed specifications for Europe's packet-switched network. These standards reduce the transaction costs of equipment purchase and, as suggested by the first hypothesis above, create a potential market for data communication equipment. In the US, regulators became involved in enforcing user preferences for a decentralized market structure, the fourth hypothesis about determinants of market structure. While AT&T sought to control the data communication equipment market by offering its own data communication equipment to customers, competing equipment producers and customers sought, and eventually won, regulatory decisions that would allow the interconnection of technically 'compliant' equipment (equipment meeting standards designed to prevent harm to the network or other users' services). Regulatory decisions supporting interconnection have continued to the present, with AT&T and now the Regional Bell Operating Companies (RBOCs) and Local Exchange Carriers (LECs) being required to interconnect competing companies or users at virtually any node within the telecommunication network. These developments have had a very favourable outcome on the variety of services and equipment that is available to meet user needs.

At the same time, because regulation has compelled a particular economic outcome, it is difficult to conclude what the advantages of vertical integration and endogenous improvements in coordination or managerial capabilities would have been if AT&T had won, rather than lost, the regulatory battles over interconnection. In particular, the current US data telecommunication system is comprised of an enormous number of mutually incompatible subsystems. While these systems may be expanded or reconfigured, the transaction costs of interconnecting them are relatively high and are a barrier to vertical disintegration of corporate information processing (David and Foray, 1994). Discussions of 'National Information Infrastructure' policies have emphasized the desirability of establishing new standards for interorganizational data communication as well as delivering new services using higher capacity data communication channels.

The advent of fibre optic networks will permit a dramatic expansion in the availability of higher capacity channels for data communication. At present, the capacity of these channels is a severe bottleneck in the delivery of a wide variety of services. We may therefore expect a new data communication infrastructure to emerge to serve the demand for higher capacity. The extent of use of this capacity will depend upon the effectiveness of the technical compatibility standards, complementary hardware and software and the prices that are charged for using the capacity. Technical compatibility standards and complementary hardware and software are both subject to the market versus non-market trade-off suggested by the Coasian framework. As noted below, however, the advantages of variety and specialization seem to favour a market rather than an internal solution to this trade-off.

We may learn something about the interaction of the factors determining capacity utilization from examining the recent experience in the use of higher capacity data communication channels in the US and Europe through the development of 'enhanced leased line' and 'publicly switched packet network' services. Enhanced leased line and publicly switched packet data networks have created markets for specialized Wide Area Networks (WANs) that are major breaches in the technological uniformity of POTS-based telecommunication networks. They are, at the same time, the first steps toward realization of the 'digital information network infrastructure'. Very rapid technological progress in the equipment markets for WANs has made it possible for sophisticated users to construct their own high-capacity data networks that integrate the flow of data from slower speed lines, local area networks and communication between mainframe computers to provide user defined services of substantial complexity and variety.[1]

Enhanced leased lines, which are available throughout industrialized countries but have a particular importance in US markets, offer a technologically differentiated service. In particular, an enhanced leased line such as one based on the T-1 standard supports a channel of higher speed data communication connecting two predefined termini. The message traffic over this channel may contain mixes of voice or data, allowing the direct connection of an organization's internal computer and private branch exchange networks. By providing high quality, assured and uninterrupted point-to-point data communication services, leased lines, in principle, serve as a bypass of the publicly switched telecommunication network.[2] The technical compatibility standards that have been defined for enhanced leased lines have created an entire industry devoted to high performance data communication equipment and the construction of WANs. By facilitating the entry of private, custom-designed WANs, technical compatibility standards have intensified the competitive pressures on publicly switched networks.

Among the most significant features of leased lines in their use in WANs is their 'customizability'. The freedom to define channel use has allowed users to create solutions that are particular to their needs and has supported competition among the suppliers of the equipment utilized to create new solutions, a direct application of the fourth hypothesis discussed above. The consequence of equipment supplier competition has been a very rapid rate of technical

progress and an enormous variety of solutions offered for the use of enhanced leased line capabilities. This market outcome suggests an important
lesson for the use of fibre optic network capability. At first glance, it appears
desirable to predefine technical compatibility standards so that all users may
use the fibre optic network, gain from the positive network externalities of
interconnection and avoid the negative externalities of congestion and information overload. However, a major result of defining such standards may be
to suppress variety and the competitive selection process that might provide
users with economic value that would offset smaller positive externalities
from universal connectivity. Moreover, as noted earlier, users are likely to
resist such standards if they undercut competitive advantages achieved by
endogenous improvements in the *use* of existing technologies. Overcoming
such resistance by arguing for adoption solely on the basis of technological
efficiency may be fruitless. It is particularly important to recognize that the
technical community's focus on technical efficiency may direct attention
away from the efficiency gains available from diversity and competition and
the motives for adopting new technologies (David and Steinmueller, 1990).

Public-switched packet data communication networks are an early example of efforts to begin the definition of new 'data highway' services by
central telecommunication authorities. They share many features of enhanced
leased line networks but differ in one important aspect. While enhanced
leased line networks are generally defined within a single company or between a company and closely related (and large) suppliers and customers,
public-switched packet networks can immediately serve inter-company data
communication needs. Like enhanced leased lines, packet-switched networks
define markets for third party equipment supply. Unlike WANs defined using
enhanced leased lines, the packet switched network relies on the use of a
single predefined structure for data communication, the X.25 standard. The
X.25 standard is a modest restriction on the use of channel capacity since it
defines only the lowest levels of connection and assembly of information into
packets. The X.25 public packet switched network in the European Union as
a whole is substantially larger than any single private packet network in the
US.[3] It is, nevertheless, unclear whether packet switching networks, despite
their larger size and uniform compatibility standard, have generated larger
network externalities than mutually incompatible leased lined WANs. One
important reason for the modest utilization of capacity in packet-switched
networks may be the suppression of variety in network connections. It is also
true, however, that the tariffs (prices) differ between packet and leased line
networks, that the larger size of the installed base of leased line equipment
may account for lower equipment prices, and that mutual incompatibilities in
systems using the packet-switched network may limit the available positive
externalities. Despite these qualifications, the more modest utilization of

packet-switching in Europe compared with enhanced leased line data communication (many of which use packet methods within private networks) in the US suggests caution in drawing the conclusion that a uniform technical compatibility standard is the best approach for assuring capacity utilization in advanced data communication networks.

If caution is warranted in the definition of a universal data communication technical compatibility standard, what approach should be considered as an alternative? One method of preserving variety in the use of improved channel capacity that can be implemented by producers or regulatory authorities is to use a 'common carrier' approach in defining the channel. In data communication, a common carrier approach mimics the leased line approach; the only technical compatibility standards that are imposed are those that prevent interference or negative effects on other users. Tariffs are determined according to the capacity utilized rather than what the capacity is employed to carry. A common carrier approach to transmitting information over telecommunication networks without intervention or involvement follows in the long tradition of physical conveyance of goods by canal, rail and other transportation networks, and the use of the POTS network for data communication. One of the main virtues of the common carrier approach stems from the minimal involvement of the carrier with what is being carried; seen from this perspective, it is, in effect, a regulatory structure specifically designed to reduce the scope for price discrimination based on the content of what is being conveyed, and to prevent the transportation provider from 'leveraging' its natural monopoly to extract economic rents on other, complementary, activities. As a firm strategy, the common carrier approach recognizes that gains from capacity utilization may outweigh gains from price discrimination or market power.

While the common carrier approach reduces the opportunities for carrier price discrimination in the services offered by third parties, it also prevents the carrier from undertaking effective partnerships with these third parties to enhance the delivery of specific types of service. The disadvantage of the common carrier approach to data telecommunication on this latter score is most apparent when one considers the possibilities for extending data networks in order to realize greater network externalities. Without carrier imposed standardization of message formats, and so on, coordination among using organizations must be achieved through market directed competition among the value added services offered by third party vendors. In short, the problems of adopting a 'common carrier' approach are ones of coordination. What could be achieved through internal coordination by a telecommunication provider must be achieved through market mechanisms. In evaluating the social welfare of the common carrier approach with the use of uniformly defined data communication technical compatibility standards, one must evaluate the trade-off between the efficiencies arising from variety with those arising from the positive externalities of the uniform technical compatibility standard. This evaluation is extremely difficult to perform *ex ante*, but the size of gains available from the common carrier approach will be enhanced by innovations that reduce its costs and increase its benefits, as discussed in the next section.

Evaluating the trade-offs between a common carrier approach and uniformly defined technical compatibility standards is even more important and intractable in the case of 'information networks'. The term 'information network' is used here to include a variety of service activities that are known by more commonly recognized terms and providers including business user information services (such as Dialog and Lexus), consumer information utilities (such as Prodigy, Compuserve and Minitel) and the growing array of more specialized services operating within the current constraints of the switched telephone network (such as fax-back providers and customer service networks employing voice mail). At present, these information service providers are organized around a modest number of technical compatibility standards for using POTS circuits including computer modems, fax machines and ordinary voice channels. Fibre optics will enhance the capabilities to offer new services and extend the variety of equipment for delivering information services. In an earlier work, I have introduced several concepts that are useful for analysing the network delivery of information (Steinmueller, 1992).

Three issues that are directly relevant to the problem of employing information networks to fully utilize fibre optics data communication capacity are considered here:

1. the comparative value of telecommunication as a distribution medium;
2. the definition of user interfaces for accessing information within a network; and,
3. the social welfare implications of imposing a uniform user interface for using this capacity.

A paradox facing information providers is that the technology that will improve the capacity of telecommunication networks will also improve other information distribution methods. Fibre optics networks are possible because of the three inventions – the semiconductor laser, the fibre optic transmission medium and the solid state photonic detector. Fibre optics will substantially reduce information transmission times, but the delivery of information using telecommunication channels will have strong competition from alternative media, such as compact discs, for the foreseeable future. Considerable effort will therefore need to be expended to develop information services that may be competitively delivered via digital networks.

One of the most attractive means of preserving the competitiveness of the telecommunication medium for the distribution of information is the improvement of user interfaces so that information is delivered selectively from databases that are much larger than users would find worthwhile to procure for their own use. Unfortunately, there is no generally accepted method for creating such user interfaces. From the computer's viewpoint, humans are remarkably heterogeneous in their information processing capacities and have the annoying habit of demanding information in forms that are tactile, context-dependent, and so forth. Moreover, the wide availability of computers is so recent that user interfaces are still evolving at a very rapid pace. Information service providers are only one of many commercial interests that are attempting to fashion user interfaces that will prove more engaging to users. One may be reasonably confident in predicting that experimentation with user interfaces will continue. It would almost certainly be counter-productive to standardize a single user interface for the use of the larger channel capacity implied by fibre optics technology. Thus, we may expect that growth in that capacity will be accompanied by a large and growing variety of user interfaces. From the viewpoint of attaining economies of scale, however, this variety is a problem. The selective delivery of information from databases that are larger than users would want to procure for their own use implies the existence of user interfaces that make such selection readily accessible. This problem is, of course, analogous to the problem of creating uniform technical compatibility standards for data communication channels. Variety conveys efficiencies in specialization and customization that are offset by the failure to achieve network externalities and other economies of scale.

Let us assume for the moment that regulatory authorities are willing to gamble on imposing uniform user interfaces in the hope that network externalities and economies of scale will overcome the benefits of further experimentation and variety. In this case, we may return to the problems arising from asset-specificity that have been debated in the Coasian framework discussed in the Introduction. Suppose that the regulator stipulates the user interface and opens the market to multiple suppliers of this single user interface. It seems highly likely that this solution will simply test the limits of natural monopoly in the provision of channel capacity to the user. The costs of implementing the user interface are likely to be relatively fixed. A company that is able to construct larger networks first is likely to be able to lower prices below those of its rivals and achieve dominance. Opening the market to multiple suppliers may initially create some variety in the supply of information services since these services can be used to differentiate the supply of the service package to users. Such market differentiation will, however, be limited because the common user interface will enable leader firms to imitate successful service innovations of competitors within a common framework.

In other words, meaningful variety in the delivery of information is likely to be suppressed by the definition of a common interface.

While this may seem to overstate the importance of the user interface as an homogenizing influence, it must be remembered that there will be opportunistic motivations at work in defining this interface in the first place. Once a dominant firm is established, Klein *et al*'s (1978) argument comes into play. The dominant telecommunication provider as purchaser of information services has an incentive to reduce the returns of the information supplier. The information supplier who has committed the fixed cost of constructing information has little or no recourse. Suppression of variety in both information resources and user interfaces seems inevitable, and the mechanism favouring internal production over contracting has been created by the artificial limitation of entry. Coase's argument that this outcome is in neither party's best interest is tempered by the presence of a growing monopoly in the delivery of information services where effective entry has been prohibited by regulatory decree. The resulting network will permit widespread price discrimination and may, as a result, achieve superior private returns for the service provider. Of course, under these circumstances, there are a lot of losers, including the ultimate user of the information services.

The purpose of this section has been to illustrate the significance of variety for enhancing social welfare in the use of data communication and information network services. While variety clearly has costs, the process of building markets for the enormous increase in capacity that will be possible with fibre optics suggests that the benefits of variety must also be carefully evaluated. In data communication, the variety represented by leased lines appears to have conveyed substantial benefits to users. Widespread experimentation with user interfaces suggests that a similar value to variety is being explored in the market for network information services. Thus, it appears premature to artificially limit the technical compatibility standards available for using fibre optic networks. Nevertheless, the failure to limit the variety of technical compatibility standards will also impose costs and create coordination failures. Are there mechanisms for limiting these costs of variety in the network provision of data communication and information services? I believe there are, and the next section is devoted to exploring several such mechanisms that may suggest research ideas and policy options.

PRESERVING COMPETITION: GATEWAYS, AGENTS AND COMMON DOMAIN STANDARDS

The choice between the advantages of variety and uniformity in data communication and information services appears to be mutually exclusive. Moreover, regulatory decision making can dramatically tilt the playing field toward either outcome. While the preceding section may seem to favour variety and decentralization, it is undeniable that gains are possible through uniformity and economies of scale from a social welfare viewpoint. It is not at all obvious how to avoid market failures in the development of the fibre optic networks since the full benefits of either a decentralized or centralized system are unobservable *a priori* and choosing one approach would appear to foreclose the other. Whatever choice is made is likely to create complex regulatory issues and difficult coordination problems for the division of economic activities among firms.

A major choice confronting regulators is the extent to which telecommunication providers will be allowed to control technical compatibility standards and engage in providing services. The choice of technical standards is likely best left to market forces rather than regulatory control except for the stipulation of specifications that will not directly impose costs on other users of the network. Similarly, if the possibility of predatory cross-subsidization can be limited, and to the extent that competing suppliers can be assured of interconnection access to users, there appears to be no reason to bar telecommunication providers from providing services.

Are there better alternatives than these minimalist interventions? The most attractive regulatory option is in the area of technical compatibility standards. Just as there are ample reasons for rejecting the imposition of a uniform technical compatibility standard, it is reasonable to limit private property rights in the *definition* of technical compatibility standards. The advantages provided to the user of proprietary technical compatibility standards for data communication or information services arise from the specialization or 'fit' of these standards to their needs and the ability to purchase equipment that will be operable under this standard. That 'fit' imposes negative social externalities in foreclosing the possibility of interconnection either in the flow of data or the connection of equipment. In some cases, such an externality may be preferable, as in the example of encoding bank or other transactions for security or confidentiality. In many other cases, however, the negative externality can be ameliorated through the creation of a 'gateway' (translator) that permits the interconnection of one proprietary standard to another (David, 1987; David and Bunn, 1988).

The value in creating a gateway depends upon the size of the negative externality imposed by a proprietary standard. If these externalities are large, then the costs of engineering a gateway may be repaid by those parties who suffer from the externality. Thus, gateways are an example of the first and second features of the Coasian framework discussed in the Introduction. The presence of a gateway technology extends the range of the network connection, allowing market transactions that, without the gateway, must be internally conducted or conducted with similarly equipped firms. It is also true, however, that gateway technologies may be endogenous, becoming more desirable as the size of mutually incompatible subnetworks expands, often making a link between them more valuable.

The problem with gateways is that they reduce supplier incentives to innovate and provide customized solutions to customers. At the limit, if gateways were completely efficient and costless, all supplier innovations could be immediately imitated by rivals who would not have to incur the costs of innovation. In practice, gateways are likely to be costly to create and maintain in the face of changes in proprietary standards. It would be difficult and possibly unwarranted to attempt to distinguish between changes in proprietary standards that were aimed at technical improvement from those that were aimed at frustrating gateway technologies. In many cases, however, the supplier may have a positive interest in the existence of a gateway since they may experience greater derived demand for their product as the result of the positive externalities of interconnection.

Beyond the minimalist regulatory interventions suggested above, the desirability of regulatory policies improving the prospects for gateway creation should be debated and resolved. The policies to consider could range from the publication of the technical specifications of proprietary standards used to transmit information over public networks to provisions for the public disclosure of technical information necessary to create a gateway in the event of abandonment of the product or failure of the firm. The case for regulatory involvement is that while direct losses from incompatibility are borne by the user and should therefore be handled with private contracts, externalities are imposed on parties that are not part of the immediate private transaction. Practically, however, the transactions costs of private contracts suggest that intervention by regulatory authorities may create significant social benefits.

A coordination problem among firms that may be addressed by private innovation is the problem of granting access to the proliferating variety of information services without requiring the purchase of a particular user interface. At present, many information services are sold exclusively to the providers of user services. There are two problems with this state of affairs. The first is that users who wish to use multiple information resources are forced to master multiple user interfaces. The second is that neither the information provider nor the user service company can receive revenue without recruiting the user. In practice, this situation has created a large growth industry in human information agents who specialize in the extraction of information from multiple sources. However, it is not a solution that provides a sound foundation for further rapid growth in the use of telecommunication networks for information services.

The most attractive private innovation for dealing with the coordination problem is the creation of automated agents that are given access to multiple user information services and/or information resources. At present, this innovation seems to be impeded by the desire of user service companies to create loyal customer bases, a 'strategic' plan for vertical integration. Building a large user base has many of the disadvantages of the creation of a uniform user interface. At present, user service companies are operating as if they were in the business of 'broadcasting information' (Steinmueller, 1992), which seems likely to be an inappropriate use of a medium that allows the service provider to gain substantial knowledge about the preferences and interests of the subscriber. In any case, several gains may be made by adopting an agent approach. Firstly, the agent may be constructed in such a way as to overcome transactions costs in accessing multiple user service companies (through the management of multiple 'virtual' single user accounts, if no better solution is offered). Secondly, once established, the agent may specialize according to user interests or interface types (the latter development is beginning to occur in a limited way through user information services that offer gateways to other user information services). Thirdly, the agent, not being tied to any particular information service, can develop a less partisan opinion of the relative merits of individual services. Fourthly, the providers of agents may market their services specifically to users who do not have reservations about the capture of information about their interests and preferences, an approach that is more difficult for user service companies that are striving to be national information utilities.

A final area to consider in attempting to address the coordination problems that would arise from preserving variety in the use of fibre optics data communication and information services is the possibility of common domain standards. At present, telecommunication standards are defined in multiple layers, often according to the Open Systems Interconnection (OSI) framework. Access to the lower levels is fairly straightforward and there are often competing suppliers at these levels. Opening higher levels to competition is more difficult. Supplier companies have an interest in capturing the returns from their large software investments in higher level proprietary standards as suggested in the gateway discussion above. Moreover, user information service companies have a greater interest in recruiting subscribers than in selling their information resources piecemeal as suggested in the discussion of agents. In both cases, however, the supplier companies are sacrificing potential gains from the larger network externalities that could be made available by opening interconnection access. The much larger channel capacity provided by fibre optic networks will offer many more opportunities for creating mutually incompatible data communication technical compatibility standards and information network user interfaces. Insistence on preserving control of these standards and interfaces at the level of the user is likely to create a growing fragmentation in the provision of data communication and information services. To some extent, this fragmentation may be reduced through the use of gateways and agents. In addition, however, the creation of specific interface standards allowing interconnections among companies prior to the delivery of data communication or information services to the user may prove desirable.

These standards for interconnection of information services, or 'common domain standards', would provide a means to change the division of economic activities among firms in data communication and information service businesses, permitting more specialization and higher proprietary rents in areas where the user demand was inelastic, while opening up broader markets for more elastically demanded services. Common domain standards are clearly complementary to the idea of agents as a coordinating mechanism; they serve to improve the efficiency of the agent. Since this particular channel for access can be priced independently, the data communication or information service provider can choose an optimal pattern of price discrimination with a multipart tariff on the various access methods allowed. Providing a common domain standard is, therefore, also in the supplier's interest.

CONCLUSION

This Chapter has considered the problem of the division of economic activities among firms in the data communication and information service industries in the context of widespread deployment of fibre optics networks. The major problem in deploying fibre optics will be the very high fixed costs of this system which will require high capacity utilization. Since public authorities, rather than markets, are likely to determine the timing of this deployment, there are real prospects for large social welfare losses. I have argued that there are clear disadvantages of attempting to gain capacity utilization through the use of uniform standards and a regulator-determined market organization for these industries. At the same time, I have noted that the costs of variety are real. Several mechanisms for preserving the advantages of variety have been noted including the facilitation of gateways, the use of 'third party' agents and the promotion of common domain standards. Facilitating the construction of gateways appears to be an activity in which regulators can be useful. Agents and common domain standards are private innovations which are likely to improve private and social welfare.

ACKNOWLEDGEMENTS

I am grateful to the Markle Foundation and the Center for Economic Policy Research for financial support allowing me to complete this paper and Robin Mansell for important substantive and editorial input.

NOTES

1. An interesting feature of wide area network management practice is the externalization of surge requirements on the switched telecommunication network. In order to reduce negative congestion externalities, companies construct smaller wide area network capacity than needed for surge capacity. When a surge occurs, these companies dial into the switched network to absorb the overflow. This practice is not particularly popular with central telecommunication companies who are confronted with intermittent and transient bursts of demand.
2. In practice, leased lines may be implemented through actual physical connections bypassing switching systems, or they may be 'virtual' connections that utilize switches, yet operate transparently and without intervention at a higher capacity than POTS to create uninterrupted point-to-point connectivity.
3. National X.25 networks in Europe have not yet been completely integrated and international connections within Europe are achieved through X.75 gateways.

17. Coalitions and committees: how users get involved in information technology (IT) standardization

Dominique Foray

USER KNOWLEDGE AND REQUIREMENTS

The establishment of a standard is fundamentally a coordination activity. The interplay of reputation, credibility and the ability to generate a coalition of common interests is an important determinant of the standards-setting process. This complex interplay was, until recently, characterized by the domination of major producers who successfully put together strategies aimed at internalizing the coordination benefits and controlling the content of the standard. According to David and Steinmueller (1993), there are different ways by which major producers can dominate and control the process. For example, government-mandated standards for procurement may be formulated so as to favour existing major producers. Government regulatory processes that issue standards are often vulnerable to capture by large, domestic producer interests, as these can provide the technical expertise necessary to write standards. Voluntary standards-writing organizations are subject to domination by representatives of major vendors, rather than users or minor suppliers, because the dominant producers have an advantage when it comes to undertaking the background Research and Development (R&D) and sending expert personnel to participate in the work of the committees. Coalitions of existing producers can use voluntary standards-writing processes to issue product specifications that impose cost burdens upon current rivals or potential users. This cornering of standardization bodies by the main producers implies, therefore, the *quasi* absence of users from the process (Rankine, 1992).

There are, however, at least two major reasons for thinking that cutting out the user results in losses as regards the social benefits of standardization. The first is connected to the issue of compatibility between products and the lack of standardization or the coexistence of multiple standards:[1] the users carry the burden of the costs linked to static inefficiency resulting from situations where multiple standards coexist and, therefore, incompatibilities between different networks exist. In these situations, the user can only hope for uniformity, that is, that one or other solution will prevail. The performance of the standard depends more on its monopolistic situation than on any particular technical specificity; 'it is the fact of the standard, rather than its form that matters' (Swann, 1991: p. 108). The users, therefore, have a specific rationale for taking action aimed at 'forcing' the market to move towards greater compatibility between products and integrated networks. The second reason concerns the fact that users are a decisive link in the chain of positive feedbacks: the learning-by-using mechanism, which is at the root of the dynamic evolution in the technology of a standard.

Compatibility Standards, Static Inefficiency Costs and Users

In some markets, it is possible that several incompatible standards persist over a long period. This situation is a source of inefficiency which is partly at the expense of users. For example, fragmentation in the computer industry, namely a situation in which not all computers can run the same software, causes difficulties for the user who will have fewer and lower performance complementary goods and services to choose from because the market is segmented, and less scope to work out purchasing strategies for goods tied to those standards. The standards themselves will be less satisfactory, especially if the goods concerned are interconnected. Examples of users penalized by the variety of standards are legion. The most striking concern firms using electronic and informatic networks. Accordingly, in a market such as microcomputing, where partially interchangeable products are on offer, increased compatibility raises the value of products from the user's point of view. Consider the case of the computer. All else being equal, consumers will prefer compatibility of operating systems (Berg, 1988). Increasing compatibility leads to rising demand so long as the increased value of products offsets the cost of altering them. Therefore, the users, being the main sufferers from the lack of integration and compatibility between networks, have reasons to adopt procedures aimed at 'forcing' the market to move towards more integrated networks.

Let us recall the experience of General Motors (GM) which resulted in the Manufacturing Automation Protocol (MAP) initiative. In 1980, GM, the world's biggest industrial enterprise, set up the Manufacturing Automation Protocol (MAP) Task Force to resolve the problems of communication between its 40,000 or so factory production units that were using robots and numerical controls. At this time, GM was spending over half of its automation budget on designing specific interfaces for incompatible machines. While computer prices were dropping, networking costs were soaring. For instance, a single factory typically had an average of 17 independent systems, each separately cabled. If they were to be able to communicate, special software had to be developed. A general reference model, Open Systems Interconnection (OSI), was selected, and two years later an ultimatum was issued to all suppliers, requiring them to comply with this model (Dankbaar and van Tulder, 1991; Bresnahan and Chopra, 1990).

Several user actions are expected to develop in the follow-up to MAP. For instance, information technology executives from a number of large companies including American Airlines, Du Pont, GM, Kodak, McDonnell Douglas, Merck, Motorola, 3M, Northrop and Unilever have met in an informal, unstructured group on a few occasions to exchange views regarding their common interest in helping to promote open systems. All companies agreed that they had a common need to accelerate the commercial availability of open systems based on vendor-neutral standards and enabling technologies in a proactive manner. In order to communicate this need more effectively to the vendors, the different user companies agreed on a set of standards and specifications to support their long and short-term needs.

The User as a Repository of Specific Knowledge

The second reason why users, of necessity, are involved is linked to their irreplaceable role in the process of developing a standard. The user has specific knowledge and masters those situations requiring the local application of the standard. By interacting with the producers, users engender learning-by-using mechanisms. In two different situations, standardization will prove especially responsive to the active role played by the user in the process: (a) when the standard is anticipatory, the standardization process has the features of a product development activity and a genuine learning-by-using process can get under way (von Hippel, 1988); and (b) when the object of standardization is a language, this can be worked out only with the cooperation of the users. The example of software is a particularly clear instance of the involvement of the users in the elaboration of the product: 'the optimal design of a software depends upon a flow of information from its customers' (Rosenberg, 1982; p. 139). David and Foray (1994) have pointed to the decisive role played by users in working out document standards for Electronic Data Interchange (EDI).

Conditions for Efficient Implementation by Users

Beyond this assessment of the potential impact of user involvement in the standards-setting process, emphasis must be given to the conditions required for this potential to be realized. Users must be in a position to express and organize their particular demands, for example regarding compatibility and integration, as well as their specific knowledge, for example regarding fields of application, with a view to putting them forward in a coherent form within standardization committees.

As regards the search for compatibility and integration, the key issue is probably the ability of users to form coalitions. This is the message of MAP as a user-initiated standard. GM's MAP initiative soon took off through the cascade effect on its subcontractors and because of the coalition formed by major users, both private (Eastman Kodak, McDonnell Douglas) and public (the US Department of Defense). A coalition was initiated by a major user firm which, because of its size and its position as the dominant member of an industrial subcontracting network, was able to recruit kindred spirits and to prevail on its own subcontractors to follow its lead. Furthermore, this major user was able to attract other large companies, thanks to its reputation and credible commitment effects. As suggested by Bresnahan and Chopra (1990), it is not commonly the case that buyers can reveal their concerns so dramatically by departing from the normal institutional procedures of vendor, committee or government standards-setting. However, this striking example of user involvement in standardization also had negative effects, amongst others the absence of participation by small firms (Dankbaar and van Tulder, 1991). These experiences illustrate, nevertheless, new types of collective actions that users introduce in standards-setting activities to achieve the goals of increasing competition, reducing the appropriability of proprietary standards and encouraging extremely broad interconnect capability.

As for potential interaction between producers and users allowing the standardization process to reap the benefits of learning-by-using, the key factor is probably the role played by lead-users as the agents generating positive feedbacks in the dynamic process of technologically evolving standards. Indeed, 'lead users are users whose present strong needs will become general in a marketplace months or years in the future. ... They can serve as a need-forecasting laboratory for marketing research' (von Hippel, 1985: p. 36). This ability to demonstrate future requirements is obviously impossible to replace in those standardization processes which anticipate the emergence of a market and are embodied in product development.

These conditions are complementary in that they involve the need for active coordination and concerted organization, complete with 'spokesmen' and incentives to further general commitment to institutions in a position to play the twofold role of forcing the market to integrate and taking an active part in the dynamic process of technological development in standards.

The analysis in this Chapter is organized in two stages: (a) the conditions required for a coalition of users to come into being are considered by applying the model of binary choice with externalities developed by Schelling which extends the 'prisoner's dilemma' to situations involving agents who are multiple but united by their rationality; and (b) the types of problem which may be caused by users intervening in the logic of collective decision-making within committees are examined together with the ways in which these problems may be alleviated.

A SIMPLE ECONOMIC MODEL OF USER COALITIONS

It is important to study the conditions necessary for the establishment and the viability of a coalition of users, capable of limiting the anticompetitive effects of standards-setting activities. Schelling's (1978) approach to collective behaviour facing individual binary choices in the presence of externalities provides an elegant device to understand the nature of the process of coalition building. Using this framework, it is possible to demonstrate how the pay-off structure changes. The coalition process has path-dependent properties – tipping points and multiple equilibria – and calls for some effort at social organization and active coordination to ensure the viability of the coalition.

'Public Good' Effects and 'Viable Coalitions'

The biggest difficulty in creating a coalition is probably that there is little incentive for the individual user to take part in the negotiating process when a standard is being formulated. This is because the effects of standardization, that is, compatibility and interoperability, have the nature of a public good (Kindleberger, 1983).

A standard has some attributes of a public good: '(a) the indivisibility of whatever benefits the goods provide among the separate members of the group enjoying them; and (b) the condition that every member of the group has equal access to the total quantity of the good which is made available by the group' (David, in this volume: p. 23). The first attribute means that the use of the standard by an individual in no way prevents other people from using it. In fact, VHS video cassette recorder owners gain from the compatibility of their equipment with prerecorded cassettes. This compatibility is considered to be a non-rival good. It is a purely technical characteristic: the utilization of compatibility by an agent does not physically exclude its utilization by others. As a result, there is little incentive to take part in the process – especially as it costs time and money to participate – every user will count on the others knowing that, in any case, he or she will profit from the benefits of the standards-setting activities.

This course of action is very well known in economics and can be analysed in the framework of the 'prisoner's dilemma' (Axelrod, 1984). Using

Schelling's (1978) elegant diagrams extensively, I begin with the basic definitions of a Uniform Multi-person Prisoner's Dilemma (UMPD). A UMPD
can be defined as a situation in which: (a) there are *n* people (users), each
with the same binary choice – C (for cooperation) or FR (for free riding) –
and the same pay-offs; (b) each has a preferred choice whatever the others
do, and the same choice is preferred by everybody. In our case, the preferred
choice is FR; that is, not to participate; and (c) whatever choice users make,
the greater the number of those who choose their unpreferred alternative (C),
the better off they are.

All users have an interest in the commitment of a majority of others to the
standards-setting process (C choice), but for everybody, it is more profitable
to abstain (FR), since the fruits of the (C) option have a public good nature.
As a result, there is a strong probability that the system may be trapped in an
inefficient situation. There is, however, a specific way to solve the coordination problem in the case of a UMPD (unlike a 2-agents Prisoner's Dilemma):
(d) there is a number *k*, greater than 1, so that if *k* or more individuals choose
their unpreferred alternative, they are better off than if they had chosen their
preferred alternative. However, if they number less than *k*, this is not true.

Following Schelling, *k* represents the minimum size of any coalition that
can gain by abstaining from the preferred choice. It is the smallest disciplined
group that, though resentful of the free riders, can be profitable for those who
join, though more profitable for those who abstain.

A simple graphical device has been applied by Schelling in a way that
makes this framework quite easy to understand: on a horizontal axis measured from 0 to *n*, two pay-off curves are drawn. One curve corresponds to the
preferred choice (FR); its left end is arbitrarily taken as 0, and it rises to the

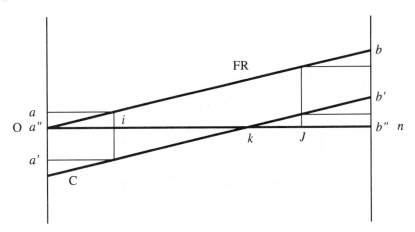

Figure 17.1   The basic representation of a UMPD

right, perhaps levelling off, but not declining. Below it we draw the curve for the unpreferred choice (C). It begins below 0, rises and crosses the axis at some point denoted by *k*. The number of users choosing the unpreferred alternative (C) is denoted by the distance of any point on the right from the left end. Figure 17.1 shows a first configuration. The only constraints on these curves are that the four extremities of the two curves must be in the vertical order shown and that the curves rise to the right and do not cross which means that in any case it is more profitable to choose FR.

At a horizontal value of *n*/3, one third of the way from left to right, the two curves show the value to a user of choosing to participate or to abstain when one third of the users choose C and two thirds choose FR.

Three propositions are important:

1. every user has an interest in the formation of a coalition, whichever choice a user makes – $b > a$, and $b' > a'$;
2. it is more profitable for everybody to abstain – $a > a'$ and $b > b'$; and,
3. in order to be viable, any coalition of users deciding to become involved must reach a certain minimum size (*k*) – $b' > b''$.

This produces a trivial result: in the presence of a public good property, the process of coalition building 'calls for some effort at social organization, some way to collectivize the choice or to strike an enforceable bargain or otherwise to restructure incentives so that people will do the opposite of what they naturally would have done' (Schelling, 1978: p. 225). We can, however, introduce additional features into the dynamics of coalition building which may facilitate concerted actions and active coordination.

Indivisibility Effects

In order to characterize a second feature of a user's coalition, not yet included in Figure 17.1, it is necessary to formulate the following question. Does the incentive to choose C – to enter the coalition – increase or decrease with the size of the coalition? In Figure 17.1, the disadvantage of the C choice is constant. This assumption misses the fact that the cost of participating in the standards-setting process will decrease as more users join the coalition. Assuming that users want to play an active role in standards-setting activities, the extreme technicality of the work in hand will put them at a disadvantage, at least at some stages of the process, since they are likely to be far less knowledgeable than the supplier. There is, therefore, a need for technical bodies to provide users with advice and support which constitutes a fixed cost. Participation in any joint undertaking is costly in terms of time and human resources and participation in standards-setting is subject to indivisibility effects (the greater the economic advantage a participant gains through the standard, the lower the relative cost of participation) and thus to economies of scale (Swann, 1991). Because of these indivisibility effects, users with a *common interest* must band together so that costs can be shared between a large number of units while each participant derives the full benefits. For example, users can share the costs of contracting with an engineering institute which will provide expertise, technical assistance and advice.

Returning to the UMPD, if joining a coalition means paying a share of the participating costs, this becomes cheaper as more users join, C benefiting more than FR from the externality. We can measure this by the change in the vertical distance between our two curves as the number of users who choose FR changes (Figure 17.2).

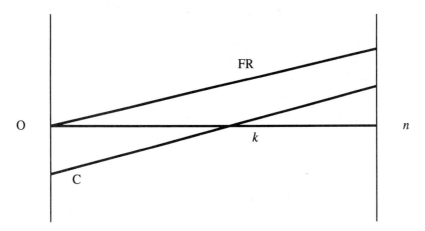

Figure 17.2   UMPD with non-constant returns

Knowledge and Information Gains: Self-Consistent Coalitions

A third feature, the ability of the members of a coalition to generate and share new knowledge, may play a critical role in giving the coalition self-consistent characteristics. Several case studies show that, in the course of the coalition-forming process, the participation (C) alternative can become the unconditionally preferred choice: beyond a number k', the coalition of users is large enough to generate some positive feedbacks based on the effects of synergetic cooperation, informational learning and the efficient use of new knowledge among the members of the coalition. As a result, the members of the coalition will derive a net profit from their commitments. A case in point was the European Research Coordination Agency (EUREKA) project, Fieldbus, including 15 participants. A study has shown that there were competitive advantages in being a member of the coalition because of learning effects and informational gains stemming from the development of pilot plans in conformity with the new standard (EUREKA Secretariat, 1993). The coalition can be defined as a club from the moment when the number of coalition members makes it possible not only to share the costs (indivisibility) but also to gain a net profit resulting from informational learning and synergies within the organization. On the other hand, users choosing to stay out do not enjoy these positive externalities based on the generation and the distribution of knowledge among the members of the coalition.

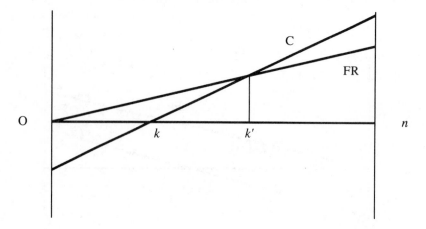

Figure 17.3   Pay-off curves in the intersecting case: the positive effect of joining the coalition

Figure 17.3 shows intersecting curves, both rising to the right: the participation (C) choice is preferred at the right and the non-participation (FR) choice is preferred at the left. Two situations of equilibrium become possible; an all-C choice and an all-FR choice. The participation (C) equilibrium enjoying externalities is the better choice. However, if everybody chooses not to participate, nobody is motivated to choose otherwise unless a large enough number of others do choose to participate to get over the hump and beyond the intersection.

Sufficiency

It is possible that beyond yet another threshold (the number k''), the coalition becomes too large: the high number of users can cause an increase in coordination costs resulting, in particular, in a slowing down of the process of interest convergence among the users and the risk of free-riding behaviour. The higher the number of members in the coalition, the higher the cost of coordination and the risk of 'free riders' taking advantage of the future standard's nature as a public good. Certain users, knowing that they will benefit from the improvements brought about by the negotiating committee, will make very little effort to contribute to the process. These two subsidiary problems tend to become less serious in the long term as the joint organization has time to get into gear. Once experience of coordination has been gained after a few negotiating rounds, organizational costs can often be reduced to practically nothing, and 'free riders' can be spotted and thrown-out. It is clear, however, that the maintenance of the coalition below a certain number is the best way to eliminate the sources of these problems[2] (David, 1991).

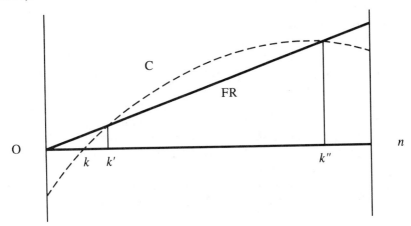

Figure 17.4   Pay-off curves, including a basic UMPD situation + non-constant returns, positive incentives and the sufficiency property

The four characteristics discussed above are illustrated in Figure 17.4 showing, with Schelling, that the generic problem does not have the precise structure of a prisoner's dilemma, but refers to all situations in which equilibria achieved by undisciplined action are inefficient. This framework produces an important result: the incentive structure is changing in the course of the process of coalition building. There are tipping points and multiple equilibria and the result of the process of coalition building is path-dependent.

Tipping Points, Equilibria and Path-dependent Properties

In a recent book, edited by Zeckhauser in honour of Thomas Schelling, Margolis (1991) shows how path-dependent properties can be derived from some of the diagrams described above. Starting with Figures 17.1 and 17.2, the only equilibrium is at the extreme left, where no one is cooperating because at every point in the figures, the FR curve lies above the C curve. As we move across the horizontal axis, everyone at some point becomes the marginal chooser, and whatever we might assume about the prevailing level of C agents, that marginal chooser always does better to free-ride. The prevailing level would move slightly to the left where another person is the marginal chooser; and that person will also choose the FR option. The only difference between the social situations described in those two figures is that in Figure 17.2, the cost of joining the coalition (V) decreases with increasing cooperation. Consequently, the cost of incentives, like subsidies to offset (V), will decrease as the fraction induced to cooperate increases.

In Figure 17.3, the strong free-riding advantage when cooperation is low decreases as we move to the right, reaches 0 at the cross-over, and, thereafter, the net advantage increasingly favours cooperation or technological club effects. In this Figure, where the C curve crosses the FR curve from below, the cross-over identifies a *tipping point* $(t = k')$, with two possible equilibria, one with 0 cooperation, the other at 100 per cent cooperation. Why is (t) a tipping point and not an equilibrium? The reason is very simple: clearly, we would always move away from (t) until we reached equilibrium, which would be at the extreme possibilities. Accordingly, the results could vary radically, contingent on how events leading to the eventual equilibrium had unfolded (see a in Figure 17.5).

If we turn to Figure 17.4, including both positive and negative incentives, we have a more complicated picture with a tipping point and one interior (not 0 or 1) equilibrium point $(Q = k'')$. Starting from any point between 0 and Q, we would always move away from (t) until we reach equilibrium (at points 0 or Q). Starting from any point between (t) and 1, the social process would always move toward Q. In this situation, as suggested by Margolis, dynamics are very important. If somehow the community of users can move to the right of the tipping point (t), it will reach the favourable equilibria at Q; otherwise, if nothing else intervenes, it will reach the perverse equilibria with 0 cooperation (Figure 17.5(b)). This tipping point configuration exhibits interesting *path-dependent properties*.

It is important to reach some points, beyond the tipping point, in order to place the process in the attractor's domain of favourable equilibria. I now explore the mechanisms and procedures of social organization, technologies

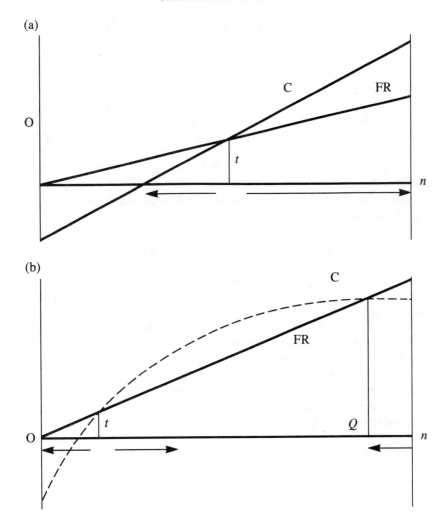

Figure 17.5   Tipping points and equilibria

and institutions that allow information technology users to develop concerted actions with respect to the coalition building problem.

BUILDING AND MAINTAINING A VIABLE USER COALITION

How can a coalition be set in train in the theoretical conditions laid out above, namely, with an incentive structure which at first incites each of the actors to adopt an expectant attitude, but which is capable of changing subsequently and encouraging the agents to behave in a more cooperative manner? The conditions which we propose are first and foremost quantitative, in accordance with Buchanan's (1965) theory, since they involve the optimal size of the users' coalition. They are also qualitative in that they involve the ways in which the knowledge generated by these user coalitions about the modes by which standards may receive specific application can be appropriated and generalized.

Quantitative Aspects: the Optimal Size of the Coalition

The problems here are twofold: (a) the coalition must be put in a position where it is capable of rapidly exceeding point (*t*), carrying out recruitment so that the dynamic process is to be found in the zone of attraction of cooperative equilibrium *Q*; and (b) procedures must be established whereby the coalition can return to the point of equilibrium if recruitment should be excessive and extend beyond Q (see Figure 5(b)).

The dynamics of agent recruitment: bandwagon, reputation and credible commitment

The creation of a viable coalition by reaching a minimum size is dependent on the interorganizational networks linking the users. Two features are of particular importance: the bandwagon effect, generated by the pre-existence of local networks of interindustrial relations, and the effect of reputation and credible commitment, generated by the actions of large users. The following cases show the importance of both features and the critical dimension of the existence of significant relational structures in the user system.

The first case deals with the key role of sophisticated major entities, for example, Boeing, GM, Electricité de France (EDF), and so on, whose information technology investment projects are on such a huge scale that they can co-opt suppliers and vendors. The example of MAP shows that it is possible for a large user to create a coalition which will reach, *quasi* instantaneously, the minimum size of viability: the large user is able to control the entire subcontracting system linked to its activity, namely the bandwagon effect. It is also able to persuade other large users to join the coalition because of its reputation and credible commitment. Bresnahan and Chopra (1990: p. 7) note:

Perhaps the most important lesson to be drawn from the factory-LAN experience, is the light these events cast on the role of the large industrial corporation. Very large firms in the relevant industries have attempted to influence the future flow of rents. These very large firms have some advantages; a bandwagon already appears to be rolling if GM is on it. A very large firm can hope to influence expectations and behaviour of both its future allies and opponents. Further, the factory environment already has an engineering and MIS department and therefore has 'informed buyer' status. This implies the existence of the buyer as the possible originator of a competing standard.

In the MAP case, the vendors rapidly felt obliged to join the coalition because of the credibility of the user's commitment.

The second case deals with the role of associative groupings, for example, professional or trade associations, which typically need to be part of a common problem-solving structure. The bar codes used in the food retailing industry are an example of the standards-setting role played by trade associations in the US. Rather than go to an existing standards-setting body, the trade associations hired the Battelle Institute to consult all concerned and to produce a final recommendation which made the advantages of using optical bar readers obvious. For the industry, the operation was an unqualified success (Rankine, 1992).

The rapid extension of a coalition towards the critical size for viability is conditioned by the existence of significant relational structures, namely the pre-existence of various clusters or subgroups of firms among whom there is a history of close industrial relations. Thus, the creation of a coalition is strongly influenced by the existence of a shared history among the potential members. Everything hinges on the structure of the network concerned.

Managing the sufficiency constraint
This problem can be addressed in two ways: firstly, it is possible to try to maintain the number of firms below a certain maximum size ($n < k''$), by increasing the membership fee or, in a more subtle manner, by establishing different levels of commitment corresponding to different types of membership. A case in point is the US MAP/TOP User Group, which offers several types of membership (Dankbaar and van Tulder, 1991). This procedure permits the co-existence of 'different coalition-sizes'; that is, keeping a large number of members in the coalition and benefiting from the efficiency of a small-size coalition which only matters in the decision process.

Another possibility ignores those mechanisms designed to discourage the entry of new members. It deals rather with the reduction of the inefficiency costs of large numbers and diversity. Through these mechanisms, it is not a matter of maintaining the size of the coalition below k'', but of moving k'' towards the right. Here, we can address the role of anticipatory standards as a mechanism which provides an important means to reduce the negative effects of diversity of agents in a large coalition. These standards result in the homogenization of agents into a few categories and allow a coalition to be viable even if it reaches a large size.

Qualitative Aspects: Institutional Arrangements Supporting a Coalition

Institutional arrangements are critical to accompanying changes in incentive structures and in allowing agents to move towards more cooperative forms of behaviour.

The provision of infratechnologies and conformance tests

The first issue involves making the expertise available to users which will help them to: (a) express their requirements; (b) check that products conform to the standard; and (c) formulate knowledge as regards specific applications. The provision of infratechnologies, that is, the set of methods, scientific and engineering data, practices and measurements necessary to elaborate the technology of the standard and to carry out testing procedures, is particularly crucial (Tassey, 1992). Infratechnologies have a high public good content and therefore require government support. Historically, vendors have developed conformance tests that have been applied internally (Weiss and Spring, 1992). Such tests do not ensure interoperability and compatibility among different proprietary standards and, as a result, users have begun to call for third party certification of conformance. These factors, including significant expansion of national capabilities in measurement, standards and testing activities, are being recognized increasingly as a critical dimension of technology policy (National Academy of Engineering, 1993).

Appropriation of new knowledge

The second issue involves the coalition's ability to evolve modes of appropriating the new knowledge generated by members of the coalition within experimental and learning-by-using processes. It is crucial that this knowledge be appropriated (a) by all the members within the coalition which can raise problems, for example, when a pilot plan is set up for a particular user; and (b) only by them.

Organization of knowledge and institutions

The final issue concerns the users' ability to organize knowledge in such a way that it can be harnessed effectively. This involves relationships between sellers and users and the need to set up institutions and transfer procedures to enable the knowledge generated by the user to be incorporated into the technology of the standard. For instance, the development of effective software is highly dependent upon user experience (Rosenberg, 1982). The modification of software systems in response to this experience is intrinsic to software engineering.

Many computer companies routinely provide extensive support that involves software modification when bugs are discovered by customers – as they inevitably are – when the software is used. Such service arrangements represent, in effect, an institutionalization of procedures for exploiting the learning-by-using phenomenon in the computer industry (Rosenberg, 1982: p. 139).

We have analysed the requirements for the setting up of a coalition and the institutional arrangements that will enable a group of users to express their needs and coherently formulate their knowledge regarding the specific application fields of the standard. It is time now to consider how increased diversity and the irruption onto the stage of a new specific interest category can be tolerated in the collective decision-making processes followed by standardization committees.

THE INVOLVEMENT OF USERS IN COMMITTEE ACTIVITIES

According to Farrell (1993: p. 1), 'an explicit cooperative way of arriving at standards is consensus or formal standardization. In this process, interested parties undertake to choose a standard by reaching explicit agreement. Even with no legal force, the agreement is likely to be a focal point, which is important because of network externalities'. The general problem of the organization of a committee can be understood on the basis of the concept of collective decision logic (Arrow, 1951): if all the participants do not have the same preference function or objective function, then there are no perfectly satisfactory choice procedures. Once interests diverge, the procedure for the choice of the standard will necessarily be suboptimal (Schmidt and Werle, (1992) provide an analysis of the CCITT). Therefore, one of the main obstacles to really effective user participation in the negotiating process is that the greater the number of participants, the slower the process is likely to be.

The Problem of Timing

One of the main conditions for efficiency in committee work has to do with the timing of the standards-setting process. The timing problem includes two features: (a) the time required to choose the 'right' standard (David, 1987; Cowan, 1991); and (b) the time it takes to achieve a design acceptable to all parties, which is a function of the diversity of the negotiating parties. The time involved is very long, for example, 6–7 years in ISO; 3 years in JTC1. This has particularly serious consequences for the speed of the work and hence for a committee's capacity to forestall the emergence of *de facto* standards. The slow and difficult genesis of a *de jure* standard often has to be hurried along under pressure from the market where bandwagon effects giving advantage to a company solution may be triggered. When the threat of bandwagon or snowball effects looms, a committee cannot carry on its discussions in serenity; it has to work quickly and diversity and representativeness tend to fall by the wayside in order to achieve consensus as quickly as possible.

In the following I shall focus on the second feature, the relation between diversity and timing, which clearly refers to the question of user involvement: the inclusion of users introduces greater diversity which must be accommodated without significantly impairing the efficiency of the process.

Some economists hold that standards generated by the complex interplay of market/committee forces are likely to yield the best results because they combine the advantages of being arrived at swiftly with the force of consensus (Farrell and Saloner, 1988). This point of view is debatable given the risk of a compromise in which quality suffers for speed's sake, of certain specifications being sacrificed and of the end result being somewhat ambiguous. This has happened in the case of EDIFACT which was launched in a mad race to compete not only with ANSI X.12 but with other standards developed by companies, and the coherence of the EDIFACT standard has suffered as a result (David and Foray, 1994).

How is it possible to circumvent this difficulty? The most obvious solution is to ensure that all the agents have a similar preference function, or believe it to be similar. This is the basic explanation for the great weakness of user participation in committees.

Let us assume that the problem of inciting users to join a committee is solved thanks to the existence of a users' coalition. It then becomes urgent to establish new organizational structures capable of allowing an increase in the diversity of the objective functions without losing too much operational efficiency. The efficiency of a committee will be measured by the speed of the process of formally establishing a standard. Two cases can be used to examine the impact of user entry on a committee's efficiency: (a) users have, or think they have, objective functions similar to those of the vendors and operators; or (b) the objective functions are different. In the first case, by definition, the arrival of users changes nothing in the collective decision logic that is already in place and there is no need for user representation. In the second case, it is necessary to find forms of organization which permit the committee to tolerate increased diversity without impairing the efficiency of the standardization process too much. Two solutions can then be envisaged.

In the first *ex ante* restriction of the field value of individual preferences, the decisive operator will be the anticipatory standard. In the second, the effects of increased diversity are attenuated by designing incomplete standards.

Anticipatory Standards and Individual Preferences

One strategy to reduce interest divergence is to produce technical designs 'in advance of the market'; that is, to establish preliminary outline specifications of what can be expected from the negotiated standard. An anticipatory standard is designed early in the process of product development and market building to avoid the formation of an installed base of, and assets specific to, local and incompatible options. It should nip any tendency toward the divergence of interests in the bud. This can induce users not to invest in a market solution, persuading them to choose the proposed prestandard instead. Producers will be dissuaded from trying to introduce their own non-standard systems and will be compelled to rally to the standard that is gradually taking shape in the committee. Thus, the prestandard has a role in forcing a convergence of actions and interests when the natural tendency is to disperse (David and Greenstein, 1990). As a kind of focusing device, it provides an important means by which the various groups of agents form a system of consistent mutual expectations.

According to Farrell (1993), standardizing in advance of the market means choosing a technology in a context of great uncertainty about what will be feasible, its costs and users' needs. Attempts to reduce uncertainty themselves cause delay as prototypes must be built and tested, customers consulted, and so on: 'this gives the process some important attributes of product development. The working of the committee will in turn resemble a research and development laboratory' (Cowan, 1991: p. 22). The anticipatory standard raises the problem of how standardization and cooperative research should be related. This problem arises for research-oriented committees working on standardization and for standardization oriented cooperative research programmes such as ESPRIT or EUREKA (Foray, 1990).

Attenuating the Effects of Diversity: Meta-standards

In addition to various institutional innovations such as new voting rules, regional levels of organization and weak property rights (Besen, 1990; Besen and Farrell, 1991; Farrell, 1992), incomplete and meta-standardization appear to play a significant role in attenuating the effects of diversity. Case studies show that the existence of a meta-standard may play a considerable role in supporting coordination by reducing uncertainty and incomplete knowledge of mutual expectations (David and Foray, 1994). Unlike anticipatory standards, defined as a means to restrict *ex ante* the scope of divergence, a meta-standard preserves the advantages of variety and allows agents to maintain some specificities as they enter the standardization process (Steinmueller, 1995). It is a mechanism designed to assist the *ex post* inclusion of agents' disparate interests and options within a unified framework. For example, the MAP specifications were based on the OSI framework which is an incompletely specified standard, a kind of large model tolerating incompatible options.

However, meta-standards do not guarantee direct compatibility between two products which conform to the same specifications. According to Spring (1991: p. 101), 'more abstract standards allowing for technological evolution lack specificity and result in products that technically conform to the standards but nevertheless fail to interoperate with other products that also conform. Perhaps the most discouraging observation is that standardized profiles, which are essentially standards of standards, may not provide enough specificity to allow for true interoperability'. OSI is a good example of a meta-standard. Its flexibility is partly due to the fact that other standards (notably IBM's System Network Architecture) existed before committees set to work. The committees' main task was to build bridges between systems based on different standards (O'Connor, 1992). OSI is simply a reference model which standardizes protocols to interconnect systems while allowing some degree of flexibility in their internal organization. It is possible to adopt any organization provided that this does not affect the protocols for external connections. The problem for users is that conformance with OSI standards does not ensure rigorous compatibility. Rather, it provides a framework for linking heterogeneous systems and there are many associations concerned with checking the conformance of standards to the reference model.

The benefits of the implementation of a meta-standard are significant. Following Farrell (1993), it is easier, cheaper and more effective to patch together compatibility through converters than in the case of competing technologies that are not constrained by a non-strict standard.[3] In this sense, the quest for universality could mean creating standards for setting local standards. At this meta-level, the universal standard is not a pure substitute designed to supplant local codes and protocols, but rather a means to support the regular formation of local standards. As demonstrated in studies of the Integrated Services Digital Network (ISDN) or EDIFACT, the main mission of a meta-standard should not be one of displacing local standards, but rather of assisting their absorption within a unified framework (David and Steinmueller, 1990; Steinmueller, 1992, 1995; David and Foray, 1994).

To reduce the delays associated with divergence of interests, it can be envisaged that, by including incompatible options, for example, the X.25 packet-switching standard, the standard remains incomplete. The standard becomes a choice model. Two products are considered compatible if they correspond to the same sets of choices on a menu. But does the inclusion of incompatible options not alter the very nature of standardization? As Farrell (1993: p. 28) has observed, the principle of incomplete standardization 'vitiates the standardization effort'. For this author, however, this is not the case; this procedure is a consequence of the difficulty of perfectly defining processes *ex ante*. Although a model does not ensure compatibility, it can help to achieve it. The same argument applies here as in the case of a meta-standard.

CONCLUSION

Two problems regarding the user's involvement in information technology standardization have been addressed in this Chapter: (a) involving users in information technology standardization requires the creation of effective user coalitions; and (b) including users in the work of standardization committees introduces more objective functions that must be accommodated. The solution to the first problem can be understood analytically and Schelling's framework has been used to show that, under certain conditions, users will 'automatically' form a coalition and under others, the coalition building process needs active coordination. Attenuating the second problem through different coordination methods can increase the expected benefits for participants in the process and thereby increase the set of conditions wherein users will 'automatically' form a coalition.

In general, from the point of view of including users, and except where there is a strong pre-existing coalition or central actor, reform of the types of standards and the degradation of aspirations toward strict compatibility will be very beneficial. The emergence of new methods of standardization, for example, meta-standards, incomplete, anticipated or gateways standards, enables one to postulate a trend which is creating disruption in the world of standardization. On the one hand, standardization offers incomplete solutions, meta-solutions and solutions 'equipped' with various sets of links and interfaces. This trend is caused by ever intensifying constraints in the management of complexity which are due to the *ex post* nature of most work on standardization. Standards must make allowances for elements of diversity and even of technological incompatibility and functional variety. This trend marks the end of the view of standardization symbolized by the tracks of a railway line where compatibility is a strictly binary affair. On the other hand, standardization attempts to anticipate market trends with a view to developing a product and working on prototypes which entails activities evermore closely linked to research and development.

The interesting feature in these developments is that both forms of activity are distinct in character and call upon increasingly different institutions and competences. The firms and public bodies concerned with standardization issues do not appear fully to have realized this fact.

NOTES

1. Compatibility is a relational attribute a product must possess to be able to be incorporated into a production, trade or consumption system. Standardization is a technology for producing *ex ante* the product attribute of compatibility. A standard is a technical specification to which products must conform in order to be compatible. The technical interface standard or compatibility standard 'provides information required to facilitate physical interactions and behavioural "transactions" at interfaces between objects, or between agents, and also between objects and agents' (David, 1987: p. 216; Cowan, 1991; David and Greenstein, 1990; Schmidt and Werle, 1993; OECD, 1991).
2. In a paper devoted to the organization of science, David (1991) studies this property of sufficiency in the case of networks of scientists.
3. If a meta-standard includes *n* standards, at most *n(n* 1) converters will be needed and one knows exactly what must be done. This is not so if there are no standards. Therefore, any restriction of options is better than none. Thanks to Robin Cowan for drawing my attention to this point.

ACKNOWLEDGEMENTS

I would like to thank in particular Robin Cowan and Thomas Schelling for meticulous comments on an earlier draft. Constructive comments by Cristiano Antonelli, Paul A. David, Georges Ferné and W. Edward Steinmueller are gratefully acknowledged. Another version of this Chapter is published as Foray (1994).

18. Standards, industrial policy and innovation

Robin Mansell

INTRODUCTION

The past decade has seen increasing recognition of standards-making and the structure and organization of standards-making institutions as important components of the innovation process. The political and economic determinants of standardization activities and their outcomes are among the many considerations that need to be taken into account in the analysis of the selection processes which lead to the development and use of new technical systems. The growing prominence of standardization activities in policy, industrial and academic circles is partly a reflection of their *strategic* implications. Standardization can influence competitiveness, trading relationships, technical design criteria and the behavioural performance of firms and public organizations.

The standards-making environment enables formal and informal networks of relationships to be established that facilitate information exchange. This environment also is one in which the boundaries between competition and cooperation among firms are negotiated. The activities of standardizers may or may not be closely aligned with short and long-term competitive strategies. Furthermore, because success or failure in agreeing and implementing standards is influenced by the wider economic and political environment, standards-making must be regarded as an important component of national systems of innovation (Freeman, 1993).

Standardization activities raise issues about the interrelationships among the institutions engaged in standardization as their primary activity and those institutions more commonly associated with industrial policy and various other modes of governance or regulation. Differences in the standardization practices among countries and regions can be expected to influence the propensity of firms to engage in trade and to promote knowledge transfer across national boundaries.

This Chapter suggests that the need to investigate the dynamics of technical design and standardization activities is growing because of their implications for the competitive and trading prospects of technology suppliers and users. Such investigation must take into account the way the complex technology design activities of standardizers interact with the marketplace and with other factors in the technology selection environment. Although a growing body of research is concerned with the impact of standards and standards-making organizations on innovation, less attention has been paid to questions about the implications of these activities for the effective formulation of industrial and trade policies. This Chapter points to several useful directions for research in this area and illustrates why these may provide a strong basis for assessing the impact of government-initiated standardization policies.

STANDARDIZATION INSTITUTIONS AND TECHNICAL CHANGE

As a result of the interdependence of standardization with factors that contribute to the innovation process – including invention, product development, design and commercialization – standards-making institutions can be conceived as providing the 'institutional glue' which links technical and institutional change. The concept of 'institutional glue' is a useful one when the aim is to develop an understanding of the implications of standardization in the information and communication technology (ICT) field. In this field there is widespread concern about the competitiveness of nationally and regionally based suppliers, the extent to which leading-edge innovation can be sustained on the basis of indigenous capabilities, and the impact of public policy measures that employ standardization as a means of achieving the faster diffusion of technologies and service applications.

The concept highlights the fact that, although the relationships between standards-making institutions and others with responsibilities for regulation, trade and industrial policy change through time, they can become 'hardened'. Decision-making procedures and technical architectures for equipment, for example, become subject to rigidities arising from the history of investment in technological designs and the weight of accumulated knowledge within a particular technological trajectory. Furthermore, the boundaries between the activities of institutions that are perceived as being competent in certain areas, for example, standards-making, can become 'set' in ways that no longer reflect the need to coordinate knowledge accumulation and the flow of information with respect to technique and design, with strategies affecting the commercial potential of products. Institutional 'hardening' also draws attention to the path-dependency of technical and institutional change. Standardization, together with a host of regulatory and trade measures, can reinforce technical designs and implementation choices in ways that reflect the history of past investment decisions rather than future opportunities associated with innovation.

Secondly, the concept draws attention to the need to consider the creative potential of institutionalized decision-making processes within standardization institutions and between such institutions and those with overlapping remits. Although much standardization activity focuses upon apparently technical issues, choices taken by participants within such organizations can involve quality considerations and a wide range of regulatory issues that are normally considered to be within the purview of regulatory, rather than standardization, policy. Technical choices which become embedded in standards can affect the prices of products and perceived opportunities for market entry and exit.

The characteristics of the interfaces between institutions with a remit for standardization and those concerned with other market-related issues change as a result of the institutional 'learning' that takes place over time, and political and economic pressures. Many factors, internal and external to the standardization institution, can contribute to the 'softening' of these interfaces. The result is often the emergence of strategic new goals on the part of firms and governments and of new priorities which create pressures for agreement on standards. As a result, the potential is created for incremental and radical innovations which alter technical trajectories and, subsequently, lead to improvements in the economic prospects of firms and nations.

The concept of 'institutional glue' is a useful heuristic device in representing the technical and non-technical factors, namely economic, political and cultural, that link standardization to industrial and innovation policy. For example, designers of ICT-based systems increasingly acknowledge that software provides the 'glue' which links computing and telecommunication hardware with the information processing requirements of users (Quintas, 1994). They recognize that technical specifications for software development and systems architectures are influenced by past practice and procedure and that system use is affected by producer-user interfaces that emerge from design and standardization activities.

Similarly, corporate strategists and public policy makers who are concerned with the competitiveness, structure and growth potential of ICT markets, recognize that the specifications for the design and interoperability of software-based networks are important aspects of organizational restructuring or 'reconfiguration'. Differences in such specifications can, for example, affect the location decisions taken by equipment and service producers and users (Tang and Mansell, 1993).

The linkages between technical design, requirements specification and standardization, and the competitive prospects of suppliers and users, are interdependent or 'glued' together in complex ways. Figure 18.1 represents

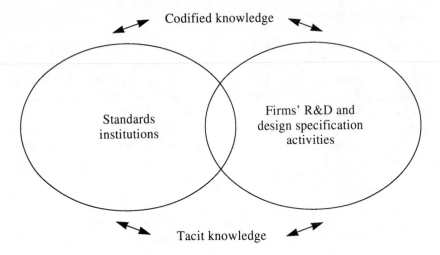

Figure 18.1   *Standardization institutions bridging R&D and design specification*

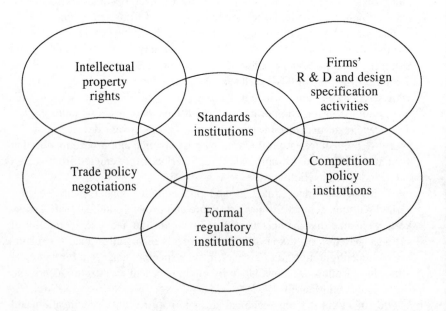

Figure 18.2   *Institutional interfaces in innovation*

standardization institutions as *interface* organizations. They provide a bridging mechanism between Research and Development (R&D) activities which shapes the design of innovative technical systems and standards and, in turn, has an impact on the take-off and viability of products in the marketplace. The interface between these domains is embodied in the flows of codified and tacit knowledge between them. However, Figure 18.2 shows that standardization institutions also interface with the activities of other organizations which are not normally considered in the analysis of the implications of standards-making. In the ICT field, for example, such interfaces are present with, for example, the General Agreement on Tariffs and Trade (GATT), the national regulatory bodies, and bodies responsible for competition policy and Intellectual Property Rights (IPR) legislation.

The optimal combination of 'solvents' that will work effectively to increase the 'malleability' of decision-making processes among these institutions is central to questions about the appropriate role of government standardization policy. The quality and nature of the codified and tacit information exchanges among such institutions must be expected to influence outcomes far beyond the purview of standardization activity *per se*. The following section considers the extent to which studies of standardization in the ICT field have focused narrowly on questions within the standards domain and, in so doing, have tended to downplay the complex factors at the interfaces between standards-making and other organizations.

PERSPECTIVES ON STANDARDIZATION

A narrowing of the boundaries of enquiry is typically a feature of any emergent field of research. It is often a reflection of a desire to deepen investigation or to apply more rigorous analytical techniques to questions of agency and structural formation (Lawson, 1993). The analysis of the standardization process is no exception. One sign of this deepening effect is the attempt to generate greater clarity as to what is meant by a 'standard'. Several definitions have come to the fore, but a dominant one in the 'standardization literature' is that given by David and Steinmueller (1993: p. 3): 'a set of technical specifications that can be adhered to by a producer, either tacitly, or in accord with some formal agreement, or in conformity with explicit regulatory authority'.

Scholars also have turned their attention to 'compatibility standards' in order to examine the way standards influence the distribution of costs and benefits of building and operating large complex technical systems such as telecommunication or computer networks as well as stand-alone technologies (Besen and Saloner, 1989; David and Greenstein, 1990; David and

Steinmueller, 1990). The problem of compatibility is examined in terms of how standardization can become 'a potent strategic instrument that can be employed to establish or further entrench a position of market dominance' (David and Steinmueller, 1993: p. 7; Tassey, 1992). Analysis of the implications of standardization proceeds by focusing on the timing of decisions to standardize and the impact that such decisions have on opportunities for market entry and the emergence of novelty and variety in technological designs.

Another line of inquiry has focused on the way distinctive producer–user interfaces are reflected in the *outcomes* of standardization conducted under different procedures of majority voting or consensus formation (Farrell and Saloner, 1988; Salter, 1988; Weiss and Sirbu, 1990). Here, consideration has been given to the way in which technology suppliers and users are involved in creating and extending coalitions in support of alternative technical standards. The cumulative 'bandwagon' effects and the costs and propensity to switch to a new technical configuration supported by competing suppliers have been investigated within the framework of decision choice models (Foray, 1993a). The question as to whether large and smaller users can be effectively involved in negotiations leading to the emergence of a standard which may then accumulate support in the marketplace has also been investigated (Hawkins, 1995b).

Yet another line of research has considered the effects of standardization on competition and the role of standards-making as a means of promoting harmonization of market-related activities in countries or regions such as the European Community (EC) (Farrell and Saloner, 1987; Hawkins, 1992b; Shurmer, 1993). Linked with these policy-oriented inquiries is work on alternative ways of institutionalizing the standardization process. Here, the advantages and disadvantages of non-market coordination, such as collaborative R&D, standardization and marketing, have been examined (Genschel and Werle, 1992; Schmidt and Werle, 1993).

In a considerable proportion of this work it has been assumed that coordination through standardization activities is strictly focused on the technical aspects of product development. Much of the focus is on whether standardization represents a mechanism that can produce efficiency gains through enhanced coordination and the reduction of transaction costs. As Schmidt (1993: p. 20) argues,

...the institutionalization of a technical orientation at the CCITT implies that relevant non-technical interests need to be translated into the common technical perspective ... This does not mean that such considerations do not influence the choice of standards, but rather that each participant needs to present the preferred option with technical arguments.

This brief review suggests that two major strands of analysis have become predominant: one focusing upon participation and advocacy within standards-making institutions; and another considering the advantages and disadvantages of markets and administrative modes of standards-making in the light of efficiency criteria and, occasionally, social or public welfare optimization (Mansell and Hawkins, 1992).

The scope of these fields of inquiry offers few opportunities to investigate the negotiation of alternative outcomes in the innovation process as a result of developments at the interface between the technical domain of standardization and the largely non-technical activities of those concerned with trade and industrial policy. For example, most of the negotiators in the GATT Uruguay Round on trade in services which aimed to increase the volume of traded services, were trade economists who regard telecommunication networks as 'electronic highways' which convey all types of services traded across national borders. Many of these negotiators argued that these 'highways' should be minimally regulated in order to afford maximum access and use (Pipe, 1993). In these negotiations, tariff policies and restrictive regulatory practices were scrutinized for their impact on trade. However, studies of how regional or national standards for new generations of network architecture would be likely to affect the tradeability of network-based services were not undertaken in this context.

Little, if any, of the tacit knowledge of those involved in the technical aspects of telecommunication network standardization filtered through the interface between standards fora and the trade negotiation fora nor was the detail of codified standards documentation regarding the interoperability and compatibility of networks treated as relevant information in the trade negotiation context. Despite their relative invisibility in the formal trade in service negotiations, however, technical design criteria and standards information affect both the innovation process and trade outcomes.

TECHNICAL DESIGN, INNOVATION AND TRADE

This section examines issues that are raised when the implications of standardization are considered in the wider context of the dynamics of institutional and technical change and with a focus on the interface between the activities of the range of governance institutions which bear on innovation and trade (Mansell, 1993). If standards are treated as the codified result of regularities of practice, then evidence of factors that contribute to the maintenance of such regularities, or create the potential for disruption in conventional practice, become the primary focus for analysis.

This approach opens the way for a broadly based analysis of the evolution of institutional structures and practices, both within standards-making institutions, and between these organizations and other relevant institutions. The structure of markets, the rise of dominant technical designs and the competitive prospects of suppliers and users, could be assessed within this framework. Technical design configurations that are embedded in ICT systems result in differential costs of supply and use and, indirectly, have an impact on the direction and speed of innovation in national and regional markets. By influencing investment trajectories, standardization influences production costs as well as whether suppliers can capture added value as a result of new product development. Industrial policy considerations come to the fore because of the capacity of suppliers to 'lock out' producers from different countries, and of governments to monitor and enforce standards which may enhance the competitiveness of home market producers and users.

Three developments help to explain the growing importance of linkages between technical design activities, standardization, innovative capabilities and trade:

1. the growing role of 'standardization' in early product design specification;
2. the increasing controversy over the appropriation of technical designs in the 'pre-standardization' phase; and
3. the strategic use of technical design to achieve the monopolization of markets.

Standardization and Design Specification

The overlap between technical design specification and standardization is clearly visible in ICTs. The deployment of new telecommunication services, for example, is almost always a collaborative undertaking between network operators, equipment suppliers and regulators. The success or failure of these initiatives in the market is often dependent on standardization agreements reached during the design and planning stages. When the telecommunication, broadcasting and cable industries were prone to many formal and informal vertical ties, there was little need to draw distinctions between 'standards' and 'design specifications'.

However, the vertical ties between network operators and equipment suppliers are weakening as a result of market liberalization measures. New entrants in equipment and service markets are proliferating. Equipment and services are now supplied with increasing frequency by non-traditional actors in the telecommunication sector, particularly by computer and software firms. The international 'club' of telecommunication monopolies and their preferred suppliers is being restructured as well. As a result, many of the 'specifications' for the interconnection and interoperation of networks are now negotiated as 'standards' among an increasingly wide range of industrial actors with varying independence from the public telecommunication operators.

Pre-standardization and Technology Development

The links between R&D and standardization in the ICT industries are increasingly evident and references are common to 'pre-standardization', a philosophy of incorporating standardization at the earliest possible phase of technical development. Although few would argue against planning for standardization requirements, much technical information in the early development phase is proprietary and difficult to coordinate except between firms that are actively involved in collaborative R&D. In the absence of procedural safeguards, 'pre-standardization' has implications for the ability of standards organizations to ensure that standards reflect a broad consensus. The reluctance, in some instances, of firms to permit access to proprietary information which has informed the development of a standard is symptomatic of tensions in the development of markets, in competitiveness strategies and in efforts to retain or build a leadership position in the innovation of new products.

These tensions have come to the fore in the European Telecommunications Standards Institute (ETSI). In March 1993, the ETSI General Assembly adopted a policy on the resolution of conflicts between standardization and the rights of intellectual property owners (Prins and Schiessl, 1993). A March 1993 'Intellectual Property Rights Policy and Undertaking' softened the original position with regard to the need for compulsory licensing for members whose designs are incorporated within standards. The Commission of the European Communities' view of such licensing had been stated earlier in 1992: compulsory licensing would be likely to reduce investment in R&D in affected sectors; non-EC firms would keep their technologies away from the EC market, and low cost equipment manufacturing entities outside the EC would benefit from cheap licences to use indigenously developed technology (CEC, 1992b: p. 5.1, 15).

A dispute is ongoing as to whether it is the Commission of the European Communities or ETSI that bears responsibility for investigating Intellectual Property Rights (IPR) claims against designs that are deemed to be 'essential' to a standard. ETSI's policy requires holders of such rights to disclose them within 180 days after a draft standard is put into an ETSI work programme. If the holder chooses not to license, and no other technical design is found, then a dispute settlement mechanism is provided. The tension between this procedure and the Commission's desire not to restrict a right holder's freedom except in exceptional circumstances is evident. Only after a 'relevant market' has been legally specified and the IPR claimant has been found to have prevented the production and marketing of a new product for which there is potential consumer demand, and to have withheld a licence in order to secure a monopoly in a derivative market, can a finding be made in favour of a challenging party.

Not only is the likelihood of a finding in favour of a challenger low because of the difficulties inherent in specifying a 'relevant market' and in demonstrating the presence or absence of potential consumer demand for innovative technologies, the time required to process such disputes can result in a *de facto* monopoly for the right holder during the period in which a technical design remains of high priority in the technical design and 'pre-standardization' process. The linkages between strategic competitiveness strategies and the technical specification of designs as candidates for standards agreement are clearly visible here although no in-depth research on the implications has been carried out as yet.

Monopolization and Standardization

The implications of standardization also can be considered in the light of theories of 'dominant design' – technical elements defining design paradigms that are sufficiently powerful to decisively influence the directions of future innovation (Dosi, 1982; Henderson and Clark, 1990; Metcalfe and Boden, 1992). The foregoing illustrates that standards can influence technical directions and commercial relations even before they are officially promulgated. Once established, these influences can be difficult to circumvent. Several major *de jure* standardization initiatives for ICT have played seminal roles in establishing technical design directions and encouraging new commercial relationships before the standards themselves have been either finished or applied (Hawkins, 1993a).

The openness of communication networks is influenced by the activities of standardizers as well as by the strategic behaviour of network operators. In both contexts, technical and organizational design considerations are paramount. Kapor (1993: p. 57), for example, has suggested that the outcome of rivalry among the network operators makes a difference to the consumer, '[but] more important are the *design principles* (both in technical architecture and public policy) under which the winning entry or entries operate'.

From a policy perspective, there is an issue as to whether intervention in the market using available alternatives, including standardization, is likely to enhance a given set of policy goals. It can be assumed, on the one hand, that a mature and fully articulated competitive market is present and that it is in this context that standardization choices are taken. On the other hand, imperfect competition, monopolistic competition or oligopolistic rivalry frequently offer more realistic ways of describing a market (Clark, 1961; Schumpeter, 1954; Shepherd, 1984). In the latter case, monopolization activities on the part of dominant firms who seek to build up, or maintain, a position of market power, are likely to be present (Clark, 1961). Such activities may include predatory and exclusionary tactics, and these are likely to embrace standardization activities.

The latter perspective characterizes the strategic behaviour of firms and emphasizes the way in which market distortions can be created through a variety of means which include the design of technical artefacts such as ICT-based networks. It draws upon theories of the dynamics of technical and institutional change to suggest that superficial signs of product differentiation, network and service competition can co-exist alongside innovative ways of achieving market closure and dominance through tactics such as the promulgation of proprietary technical designs through standardization.

Such practices can result, for example, in the increasing segmentation of the development of public networks. As one facet of the dynamics of monopolization, this line of inquiry opens fruitful avenues for further exploration of the importance of standardization in the technology selection process. For instance, in the case of the development of telecommunication networks, while the number of reciprocal agreements among operators grows with the liberalization of markets, innovations in technical design are creating opportunities for suppliers to create barriers to entry through standards-making activities (Mansell, 1993). A focus on the dynamics of monopolization allows questions about the way political and economic power are expressed through processes of innovation in technologies and institutions to be framed in a way that takes into account the interfaces among standardization and other institutions which, together, have an impact on the competitive advantage of firms and the growth potential of national (or regional) systems of innovation.

STANDARDIZATION AND INDUSTRIAL POLICY

Empirical analysis of the configuration of institutions at the national or regional levels that support innovative capabilities in the development, production and use of ICTs is needed in parallel with studies which aim to develop elegant models of the determinants of the standardization process and its outcomes. The relevant institutional configurations embrace standards organizations and their interfaces with agents responsible for related policies. Major shifts over the past decade in the balance of trade in ICTs and services in the international economy give considerable weight to the need for such analysis (EITO, 1993; OECD, 1993a).

Government standardization policy presently tends to be formulated on the basis of the need for intervention that seeks to 'direct' the activities of firms, or for reliance on the forces of the market. In the case of standardization, this translates into the issue of whether special remits to achieve harmonization of technical standards (often including issues of tariffing and network use) should be given to standards institutions. The alternative is generally to leave it to the market to produce appropriate and timely standards.

The outcomes of these alternatives are uncertain and highly unpredictable when one departs from the confines of theoretical models based on assumptions about idealized user preferences and choices. For example, choices as to when to standardize a technical design or to encourage diversity do not necessarily reflect the relative superiority of alternative design innovations. They are often the result of oligopolistic competition, political bargaining processes and conflict resolution not just within, but also among, a large number of institutions and actors with an interest in the outcomes (Hawkins, 1992a; Mansell, 1990; Utterback and Suarez, 1993). Thus, factors external to the parameters of relatively narrowly defined theoretical models often account for a significant amount of the variation in outcomes in the development of innovative technologies.

Policy research in the standardization field could make a significant contribution by examining these 'external' factors. Hawkins (1995a) has argued that standards-making is not primarily a process of coming to a consensus on how existing or future technical practices should be codified. Rather, it is one of determining whether any basis for consensus exists within a committee setting. If such a basis does exist, the issue is whether consensus can be created around particular elements of technical systems. One important set of questions of policy interest is therefore how voluntary/consensus standards negotiations embody the expectations of participants in their activities. This requires research on the way standardization fora of different types represent public and private interests in the design, production and use of technologies. The aim would be to analyse processes of decision-making in order to better understand what agents actually do and how the process of negotiation is conducted.

A complementary set of questions would extend the analysis of what economic agents do within standards-making institutions to what they do at the boundaries or interfaces between standardization bodies and institutions with remits in the competition, regulatory and trade policy area. For example, what occurs in the negotiation of new technical standards and what scope does this leave for the emergence of innovative technical designs, for new entry in the market, and for the maintenance or extension of market share for incumbent firms? Similarly, what public and private sector agents do in negotiations concerning the liberalization of markets and the imposition of new regulatory frameworks is likely to be influenced by, and to have an influence on, the timing of standards-making. This, in turn, must affect cycles in the product development process as well as the decline of earlier innovations.

Empirical research on whether initiatives by governments to enhance the transparency of standardization activities in the name of stimulating innovation and competitiveness also is needed. It may be that such initiatives have the effect of illuminating those areas of technical development that are most sensitive to competition rather than broadening the basis for cooperative standards development. The result could compromise public interests in promoting the development of open systems in the ICT domain. For example, choices in the adoption of proprietary or open standards must be expected to interact with supplier pricing strategies, the quality of products and the structural characteristics of markets; this, in turn, will have a bearing on the conditions of network access.

The specific nature of this impact is likely to remain largely speculative until research along the lines suggested here is undertaken. Research that would go some way to redress this lacuna would include studies of: (a) the way standardization fora represent public and private interests in the design, production and use of technologies; (b) the role of standardizers and other agents at the interface between standardization bodies and institutions with remits in competition, regulatory and trade policy areas; (c) the potential conflicts between government policy initiatives in the standardization area and the public interest in open systems; (d) the degree of coordination between the strategic positions taken by firms within standardization fora and in other institutions; and (e) the role of standardization as a mode of governance that interfaces with other governance institutions within national, regional and international systems of innovation.

Technical standards are closely linked to the geo-politics of technological innovation and diffusion. There need not be evidence of a direct correspondence between the individual choices and negotiating power of standardizers and the implications of overall trends in the choice of standards to assert the possibility of such a linkage. As the promulgation of standards affects the leverage of market share, it cannot be assumed that standardization fora represent an opportunity for cooperation that excludes rivalry among the participating firms. Although the relationship between strategic competition goals and the standardization process is anything but straightforward it should be recognized that individuals who participate in technical fora may also advise, or, increasingly, directly participate in non-technical fora which influence decisions on trade, competition and regulatory policy.

An outstanding issue that awaits empirical investigation is whether firms that achieve a high degree of coordination between their strategic positions in standardization fora and other regulatory, competition and trade policy fora, also exhibit greater and more sustained competitiveness. Cowhey and Aronson (1993) have argued that governments are seeking to 'regulate' in new ways and are using a variety of measures in an attempt to prise open markets to foreign competitors while simultaneously seeking to embrace protective industrial policies. It seems likely that standardization – defined to include more than consensus on a specific technical design or configuration – will emerge as a major determinant of differentiation in national or regional 'capabilities to innovate, imitate and generally exploit innovation efforts competitively' (Dosi *et al.*, 1990: p. 35).

If standardization is one of the central elements in the 'institutional glue' that binds public policy and private sector strategies and the 'learning' and experimentation that lead to the selection of new technical designs, it is essential to understand how the results resonate with the activities of other public institutions. Standards are also important in the context of corporate governance institutions. Although theories are being developed concerning the emergence of 'corporate coherence' and research is focusing on 'learning, path dependencies, technological opportunities, selection and complementary assets' (Dosi *et al.*, 1992), the differential capabilities of firms to effectively 'manage' their participation in standards-making and other related institutions have yet to be explored. For example, do different strategic approaches to participation in regional or national standardization institutions matter for competitiveness? Does standardization provide a beneficial means of acquiring knowledge about competitors' strategies through the exchange of codified and tacit knowledge?

The standardization process, like other processes of decision-making in related policy fora, is characterized by considerable uncertainty and the bounded rationality of the players in their respective activities. Nevertheless, there is evidence suggesting that in the early design of innovative products and further along in the commercialization phase, standards – and in some cases the failure to standardize – play a role which coincides with the strategic concerns of firms. Such concerns may give greater emphasis to the need to enhance coordination and compatibility among the components of large technical systems at one time, and to the need for greater differentiation to strengthen competitiveness at another. The standardization process, whether by publicly mandated committee or by market-led initiative, is not directly linked to issues of competition and monopolization, but neither is it wholly independent. Research targeted at instances where rapid technical innovation coincides with the transformation of historically monopolistic market structures to those characterized by competition could provide a useful indicator of the reciprocal spillover effects between standardization activities and other governance institutions.

CONCLUSION

This Chapter has highlighted theoretical and empirical approaches to research on the implications of standardization. Research on the institutional interfaces among a range of governance institutions in the ICT field could contribute to an understanding of how standardization activities are related to other forms of public and corporate governance which influence the innovation process. The importance of focusing on the determinants of technical and institutional change has been stressed as a means of assessing the potential synergies and conflicts in standardization outcomes and the economic and political strategies of corporate and public actors. The analysis has highlighted the need to examine the political and economic incentives created by the changing configuration of institutions which affects the innovative capabilities of technology producing and using firms. These include institutions of regulation, competition and trade policy as well as standardization organizations. A strategic perspective on monopolization has been presented as an effective way of framing research issues for future empirical investigation.

ACKNOWLEDGEMENTS

The Economic and Social Research Council's support for this research through its Programme on Information and Communication Technologies (PICT) is gratefully acknowledged as are comments on an earlier draft by Richard Hawkins and Susanne Schmidt.

19. Conclusion

Richard Lipsey and Robin Mansell

The Chapters in this volume provide substantial justification for regarding standards and the standardization process as fertile ground for future empirical and theoretical research. In their codified form, standards embody the outcomes of a wide range of choices with regard to the selection of technical innovations and socio-economic organization. The processes whereby standards are generated, and on occasion fail to emerge, are sites of continuous negotiation among a large number of actors often representing a mix of private and public interests in the development and diffusion of technologies. In both the environmental field and the information and communication technology (ICT) field, standards-making arenas provide fora for the exchange of considerable tacit knowledge concerning the advantages and disadvantages of historical practice in the development and implementation of technologies as well as the potential and opportunities for change.

The authors have brought multidisciplinary perspectives drawing upon economics, political science and sociology to bear on the determinants of standardization procedures and outcomes. Many of these have been shown to overlap with those which affect policy making and corporate decision making in areas such as R&D, regulation and trade, and they generally have not been linked explicitly with the role played by standardization activities in the economy. As a result of the preparation of the Workshop and the subsequent discussion, we have concluded that in-depth analysis of the contribution of standards and the standards-making process should be treated as an important component of studies within the wider framework of the analysis of the dynamics of technical and institutional change.

In his keynote address, Bengt-Åke Lundvall identified a major problem which will have to be overcome before 'standards' can be mobilized effectively as a research issue. This is the lack of a common language between standards makers and those who are ultimately affected by the standards they produce. He noted that, thus far, a relatively small amount of research has succeeded in demonstrating that standards are an important phenomenon. If further research is to result in mutual benefits for both developers and users of standards, standards-making will need to become divested of its arcane image and be directly integrated into the prevailing discourses of political and economic life.

Identifying the means for such an integration was offered as a challenge to the Workshop delegates. In the final analysis, three days of discussion did not yield definitive answers to the many questions surrounding the standards issue. However, it did succeed in exploding some myths and questioning assumptions about what we do and do not know in this area.

In his concluding summary, Workshop Chairman, Richard Lipsey, made a number of observations highlighting major themes which had emerged as a result of the presentations and discussions and which pointed to potentially productive new areas of research. These are summarized below.

PERSPECTIVES ON THE ROLE OF THE 'USER'

It became clear as the workshop proceeded that there is as yet no common understanding of the underlying relationship between technology users, the process of standards-making and the standards which emerge. The participants in the Workshop had a wide variety of perspectives on the role of 'users'. Some argued that 'users' could not, or should not, be directly engaged in the technical aspects of standards-making. Others took the view that this perspective was condescending and a reflection of the strong 'technology-push' orientation which historically had characterized certain industry sectors and, especially, the telecommunication industry. In the 1990s, in the telecommunication sector, as well as in the information and communication industries which have tended to be more responsive to market demand, users were characterized as having an important role to play. Nevertheless, although it was acknowledged that 'users' must have some influence in the development of standards if user–producer relationships are to be strengthened, it was suggested that the intimate involvement of users could be expected to result in sometimes counterproductive and even unexpected results.

Research oriented towards investigating the constraints in terms of time, money, expertise and priorities, which confront technology users who seek to influence the standards making process, could be particularly beneficial in contributing to a better understanding of the disparate expectations of producers and users of advanced technological systems.

THE INDUSTRY-SPECIFIC IMPACT OF STANDARDIZATION

Taken together, the Workshop papers and subsequent discussion also highlighted the fact that studies of standards have tended to concentrate to a considerable degree upon individual cases of the promulgation or implementation of a specific standard or protocol or on abstract theoretical models which illustrate game-theoretic alternatives under given market conditions. In the former case, significant generalization is not yet possible and, in the latter, there is considerable evidence to suggest that the implications of standards are highly industry-specific. Thus, for example, if the question is asked whether the market is likely to produce too much or too little standardization, the issue is likely to be posed too broadly to admit a clear answer with regard to the timing and impact of the emergence of a particular standard in the light of developments in the marketplace.

Research by scholars, including Paul David (Chapter 3), has contributed to our understanding of the path-dependency of technical change and innovation. Market incentives are rarely unguided by various forms of public sector intervention, and theory makes no predictions as to whether the unaided market may be expected to produce too many or too few standards. Any assessment of the relative merits of standards which does emerge as a result of the interplay between markets, standardization activities and a host of public sector initiatives must depend on the specific characteristics of the industry that is the focus of analysis. The nature and dynamics of industry-specific innovation processes, the nature of the decision-making bodies and uncertainties as to the appropriate goals and outcomes of standards-making can combine to produce very different outcomes. This is evident when the environmental field is compared to the ICT field and differences are also likely to be found in other sectors which make considerable use of standardization and codes.

Richard Lipsey pointed to the fact that studies such as those by Stanley Besen (Chapter 13) have shown that the adoption or non-adoption of standards may have different implications depending on the outcome of highly differentiated competitive processes. For example, if firms are competing with respect to a particular technology there are four possible options: (a) a firm may choose to adopt its own technical design or standard; (b) it may choose to adopt a competitor's standard; (c) competing firms may choose not to seek any consensus on compatibility standards; or (d) a consensus may be reached on the adoption of a standard which does not follow those being developed by any of the players in the market.

The priority given to each of these options will depend on the nature of the innovation process and the benefits of any single option at a given point in time will depend on the interests and incentives facing firms as they compete in the marketplace. The optimal choices with regard to the development and adoption of standards will also be influenced by the stage of innovation. Firms will be likely to have different perceptions of the benefits and costs of alternative choices at the early precompetitive design stage, as compared to the first phases of commercialization or when technologies and markets have reached the point of maturity.

Choices as to the timing of standardization and assessments of the economic impact of the adoption of a standard are also likely to be significantly influenced by the characteristics of the decision-making body. Research has suggested that the negotiation of standards is not 'technology-neutral' and that it is influenced by the process whereby decisions on standards are reached. As Liora Salter has suggested (Chapter 4), research in this area must recognize that uncertainty often pervades the individual and collective perspectives of those engaged in the standards-making process. For example, in some cases, standardizers are not fully aware of the wider implications of their activities, they may be unclear as to the types of signals they wish to send to the market as a result of their activities in the technical domain, and there may be a range of activities underway which create 'noise' so that no clear optimal choice for a standard emerges or is apparent to those engaged in the negotiation process.

The result may be a situation in which it is difficult to predict the outcome of any given standardization initiative. It may be unclear what the 'interests' of the technology producers are and whether technology developers' perspectives coincide with broader strategic concerns. In short, the evidence suggests that, insofar as standards-making bodies are unclear as to their own objectives, there will be considerable 'noise' in the development process, thus making the clear prediction of outcomes over time extremely difficult.

Although the fact that standards play a role as a strategic tool in interfirm competition was discussed by many of the Workshop participants, less attention was paid to the role of standards as a competitive tool among nations. Referring to the 'battle among nations', Professor Lipsey pointed to a hypothetical circumstance in which technology 'q' is being developed and country 'j' determines that it can capture a substantial share of the world market regardless of the standard that is adopted. Country 'j' adopts a standard that appears to be emerging and engages in production. Country 'f', however, worries about its home-based producers and is unsure as to whether they can stand up to the competition from country 'j'. Country 'f', therefore, resists the adoption of universal standards, in the hope that, sooner or later, country 'j' firms will have their progress delayed by having to convert to a country 'f' standard or that a part of the world-market will adopt this standard giving country 'f' firms some advantage over country 'j' firms. In effect, a game ensues whereby countries and firms attempt to protect some part of a market by resisting the adoption of universal standards. These facets – the timing of standardization and choice of standard – can have a substantial impact on the competitive prospects of firms and the trading relationships between different countries. As yet, research has not produced a practical description of a 'standards life cycle' which could be examined in the light of the life cycle of product development or the emergence of new markets.

Standards and the standardization process also can interact with other aspects of government policy and this must be taken into account in any assessment of the likely impact of alternative choices with regard to a given technological configuration. In consequence, standards issues must not be considered in isolation. From a research perspective, there is a need to consider the range of policies that is likely to encourage innovation and technical change and to investigate how policies with regard to standards fit as one part of an integrated constellation of policies.

STANDARDS, INNOVATION AND THE DIFFUSION OF TECHNOLOGIES

The Workshop discussions pointed to the need not only to understand the impact of standards and standardization on the innovation process, but also on patterns of technology diffusion. Bengt-Åke Lundvall and Robin Mansell (Chapters 2 and 18, respectively) stressed that standardization activities should be treated as the core of the innovation process, but that this raised the question as to whether equal attention should be given to developments at the 'technological frontier' and to those factors which play a role when technical systems are available for commercialization on the market. On the basis of available knowledge, it seems clear that the factors contributing to both the innovation and diffusion process are important. This is especially the case where the widespread diffusion of technologies has been shown to be a slow and costly process.

Although standards are clearly important factors in the emergence of innovative technical systems, the protagonists in the Workshop debate were unable to establish just how central they are to the processes of innovation, technical change, national competitiveness and economic growth. This is clearly an area that would benefit from further research. Work in this area would also need to give more considered attention to the appropriate role of public policies which may be instrumental in shaping standards development, in supporting or subsidizing the participation of actors in standardization activities, and in setting the trajectories for technical change, competitiveness and economic growth and development.

BENCHMARKING THE EFFICIENCY OF STANDARDS MAKING

Policy research in the standardization field has yet to develop a methodology for benchmarking or realistically assessing the efficiency of the standards-making process. This issue was shown to be considerably broader than that addressed within the traditional confines of economic theory where the concept of 'efficiency' is defined in terms of static relationships in a production process characterized by alternative inputs and outputs. The benchmarking problem is related to the assessment of the merits of one system of organization of standardization – a 'good system' – as compared to another – 'bad system'. Research is needed that focuses on the elements which contribute to the perception that a 'good system' is operating in any given context. It is also necessary to recognize that standardization activities occur through a dynamic process of decision-making which departs considerably from the economists' static world. Under changing technical and market conditions there is a need to assess the kinds of organizational and related changes that might be expected to result in outcomes associated with a 'good system'.

In the course of the Workshop a number of parameters that might be expected to characterize alternative systems emerged. In some cases, for example, a 'good system' of standardization might be oriented towards performance objectives rather than towards the specification of technical details and architectures. Although the primary aim of standardization initiatives is generally to encourage effective competition, rather than the monopolization of markets, standards-making systems may need to be flexible to encourage innovation or they may need to proceed rapidly to provide a foundation for re-innovation and the diffusion of new products. While it may at one time be desirable to avoid 'lock-in' to a particular technical configuration, an agreed or unified standard at another time may provide the basis upon which to focus further iterations in development activity.

STANDARDIZATION AND THE INSTITUTIONAL CONTEXT

Given the time dependency of standardization activities, it is clear that there is no single optimum solution or set of trade-offs that will produce an optimum result. Under closely defined static conditions, economic theory would seek to pinpoint a single peaked function indicating the combination of attributes (timing, diversity versus agreed standard, and so on) that would produce an optimum outcome in terms of the rate of diffusion and profitability of a given product. On the basis of research on the standardization process, it is more likely to be the case that there will be a consensus on the attributes of very 'bad' standards-making procedures, that is, a highly inflexible process rendering a negative pay-off to its participants. However, it is also likely that the range over which a flat pay-off function exists, namely the mix of characteristics and likely positive outcomes, will be extremely broad. Under these circumstances no clear signals as to the 'efficiency' of a given standards-making system are likely to emerge. Insufficient research has been carried out thus far on whether highly differentiated standards-making institutions and procedures produce broadly acceptable results, and whether there are some combinations of characteristics of the standards-making process that do tend to result in negative outcomes. While it is possible to theorize about likely outcomes under given parameters, this area calls for in-depth empirical research on the actual process of standardization and the way in which it is influenced by factors such as personality and leadership, cultural attributes, as well as changing political and economic priorities.

It is clear that the goals and objectives as well as the process whereby standards decisions are reached are of equal importance. The perceived value of decisions reached as a result of a given process which deviates to a greater or lesser degree from 'best practice' may need to be modified – but there are few indications as to what the optimal combination of features of an improved decision-making process would be. At this stage in the development of the standardization field as an area of inquiry there is insufficient evidence to provide a basis for judging which processes are superior in terms either of narrowly defined efficiency criteria or of broader criteria such as whether they generate perceived social, political or economic benefit for a particular constituency.

STANDARDIZATION AND TRADE

One key area which was not addressed to any great extent during the Workshop was the question of the importance of standards, standardization processes and trade. On the one hand, standards can play a role in encouraging trade and, on the other, they can perform a role as non-tariff barriers to trade. Standards which are used as non-tariff barriers to trade are usually related to 'attributes' as opposed to 'performance' criteria. Typically, a non-tariff barrier defines a standard with respect to an attribute (for example, wood is inflammable and therefore cannot be exported) as compared to performance criteria (for example, the extent to which wood must be treated and rendered fire proof). In the latter case, a product can be altered or treated so that it meets performance standards. In the former, it simply cannot be exported because of its intrinsic characteristics. The extent to which such distinctions are drawn and employed in the technical standardization arena where competition for markets for innovative products and services is at stake has yet to be explored in the international context. In-depth work in this area has received a considerable stimulus in the context of the negotiation of the Canada–United States Free Trade Agreement and is likely to continue to generate interest in the European Union as well as other regions where standards come to be regarded as potential non-tariff barriers to trade.

In summary, although some argue that 'standards' issues lie at the core of the innovation process, the important outstanding questions for policy research concern the role of standards-making in the innovation process and its implications for technical change, the diffusion of products, the competitiveness of firms and countries, and economic growth and development – together with the appropriate role for public policy. For the policy research community there are two main challenges: firstly, the development of conceptual approaches that will enable further investigation of the practice and outcome of standards-making in a more coherent way; and secondly, the application of this approach to achieve an integration of multidisciplinary theoretical approaches which focus on the dynamics of technical and institutional change with the empirical reality that confronts technology producers, users and policy makers.

Our collective goal should be to ensure that clarification of the determinants of the sometimes 'arcane' field of standardization helps to shed new light on the ways technical change, public policy and private investment give rise to new opportunities for the growth and development of economies in ways that are socially and environmentally acceptable.

NA ⟶

Bibliography

Abelson, P.H. (1993), 'Editorial: regulatory costs', *Science*, **258**, (8), 159.

Adams, C.A. (1956), 'The National Standards Movement: its Evolution and Future', in D. Reck (ed.), *National Standards in a Modern Economy*, New York: Harper & Brothers.

Agar, M.H. (1981), *The Professional Stranger: an Informal Introduction to Ethnography*, New York: Academic Press.

Alberthal, L. *et al.* (1993), 'Will architecture win the technology wars?', *Harvard Business Review*, May–June, 162–70.

Allen, D. and Gilbert, J. (1993), 'Standards and Convergence, New Realities', mimeo, Pacific Telecommunication Conference, Honolulu, January.

Anderson, E.S. (1991), 'Techno-economic Paradigm and Typical Interface between Producers and Users', *Journal of Evolutionary Economics*, **1**, 119–44.

Angel, M. (1994 forthcoming), 'Critique de la valeur d'existence', *Revue Economique*.

Arrow, K. (1951), 'Alternative Approaches to the Theory of Choice in Risk-taking Situations', *Econometrica*, **17**, 404–37.

Arthur, W.B. (1989), 'Competing Technologies, Increasing Returns and Lock-in by Historical Events', *Economic Journal*, **99**, March, 116–31.

Arthur, W.B. (1988), 'Competing Technologies: an Overview', in G. Dosi, C. Freeman, R. Nelson *et al.* (eds), *Technical Change and Economic Theory*, London: Pinter Publishers.

Australian CCITT Committee (1992), 'Consumer and User Participation in Telecommunications Standards Setting', Document ITSC3 (92)–34, Interregional Telecommunications Standards Conference, Tokyo, 5–6 November.

Axelrod, R. (1984), *The Evolution of Cooperation*, New York: Basic Books.

Barry, A. (1990), 'Technical Harmonisation as a Political Project', in G. Locksley (ed.), *The Single European Market and the Information and Communication Technologies*, London: Belhaven.

Becker, G.S. (1983), 'A Theory of Competition among Pressure Groups for Political Influence', *Quarterly Journal of Economics*, **98**, (3), 371–400.

Bennett, G. (ed.) (1991), *Air Pollution Control in the European Community: Implementation of the EC Directives in the Twelve Member States*, London: Graham & Trotman.

Berg, S. (1989), 'The Production of Compatibility: Technical Standards as Collective Goods', *Kyklos*, **42**, (3), 361–83.

Berg, S.(1988), 'Duopoly Compatibility Standards with Partial Cooperation and Standards Leadership', *Information Economics and Policy*, **3**, (1), 35–53.

Besen, S.M. (1992), 'AM versus FM: the Battle of the Bands', *Industrial and Corporate Change*, **1**, 375–96.

Besen, S.M. (1990), 'The European Telecommunications Standards Institute: a Preliminary Analysis', *Telecommunications Policy*, December, 521–30.

Besen, S.M. and Farrell, J. (1994), 'Choosing How to Compete: Strategies and Tactics in Standardization', *Journal of Economic Perspectives*, **8**, (2), 117–32.

Besen, S.M. and Farrell, J. (1991), 'The Role of ITU in Standardization: Preeminence, Impotence, or Rubber Stamp?', *Telecommunications Policy*, **15**, (4), 311–21.

Besen, S.M. and Johnson, L.L. (1986), *Compatibility Standards, Competition, and Innovation in the Broadcasting Industry*, Rand Corporation, R-3453-NSF.

Besen, S.M. and Saloner, G. (1989), 'The Economics of Telecommunications Standards', in R.W. Crandall and K. Flamm (eds), *Changing the Rules: Technological Change, International Competition, and Regulation in Communications*, Washington: The Brookings Institution.

Boehmer-Christiansen, S. and Skea, J. (1991), *Acid Politics: Environmental and Energy Policies in Britain and Germany*, London: Belhaven.

Boehmer-Christiansen, S. and Weidner, H. (1992), *Catalyst versus Lean Burn*, Berlin: Wissenschaftzentrum, FS II 92–304.

Bohm, P. and Russell, C.S. (1985), 'Comparative Analysis of Alternative Policy Instruments', in A.V. Kneese and J.L. Sweenay (eds), *The Handbook of Natural Resource and Energy Economics*, Amsterdam: North-Holland.

Brady, R.A. (1933), *The Rationalization Movement in German Industry: A Study in the Evolution of Economic Planning*, Berkeley: University of California Press.

Bresnahan, T. and Chopra, A. (1990), 'User's Role in Standard Setting: the Local Area Network Industry', *Economics of Innovation and New Technology*, **1**, (1/2), 97–110.

Brusco, S. and Cottica, A. (1993), 'The Economics of Green Consumerism', 2nd International Conference of the Greening of Industry Network, Designing the Sustainable Enterprise, Boston: 14–16 November.

Buchanan, J. (1968), *The Demand and Supply of Public Goods*, Chicago: Rand McNally & Company.

Buchanan, J. (1965), 'An Economic Theory of Clubs', *Economica*, **11**, 1–14 February.

Buchanan, J. and G. Tullock (1962), *The Calculus of Consent: Logical Foun-dations of Constitutional Democracy*, Ann Arbor: University of Michigan Press.

Burgess, R. (1981), *Field Research: A Sourcebook and Field Manual*, London: Allen & Unwin.

Burgess, R. (1984), *In the Field*, London: Allen & Unwin.

Cargill, C. (1989), *Information Technology Standardization: Theory, Process, and Organizations*. (n.p.) Digital Press.

Carlton, D.W. and Klamer, J.M. (1983), 'The Need for Coordination Among Firms, with Special Reference to Network Industries', *The University of Chicago Law Review*, **50**, 446–65.

Carnegie Commission on Science, Technology and Government (1993), *Risk and the Environment: Improving Regulatory Decision Making*, New York: Carnegie Commission.

Carver, B. (1992), 'Open Government', *Networking*, May, 81–82, 86, 89.

Cave, M. and Shurmer, M. (1991), 'Standardization Issues for HDTV', in J.P. Chamoux (ed.), *Deregulating Regulators? Communications Policies for the 90s*, Amsterdam: IOS Press.

CEN/CENELEC/ETSI, 'European Standardization' Brussels, n.d.

Cicourel, A.V. (1973), *Cognitive Sociology*, Harmondsworth: Penguin.

Clark, D. (1993), 'Novell will Transfer Trademark Rights for Unix to Industry Consortium', *Wall Street Journal*, 12 October, B6.

Clark, J.M. (1961), *Competition as a Dynamic Process*, New York: The Brookings Institution.

Coase, R.H. (1972), 'Industrial Organization: a Proposal for Research', in V.R. Fuchs (ed.), *Policy Issues and Research Opportunities in Industrial Organization*, New York: National Bureau of Economic Research.

Coase, R.H. (1937), 'The Nature of the Firm,' *Economica*, **4**, pp. as reprinted in O.E. Williamson and S.G. Winter (1991), (eds), *The Nature of the Firm: Origins, Evolution, and Development*, Oxford: Oxford University Press.

Codding, G.A. (1991), 'Evolution of the ITU', *Telecommunications Policy*, August, 271–85.

Coleman, W.D. and Skogstad, G. (1990), *Policy Communities and Public Policy in Canada*, Toronto: Copp Clark Pitman.

Commission of the European Communities (CEC 1992a), *Towards Sustainability: a European Community Programme of Policy and Action in Relation to the Environment and Sustainable Development*, COM, (92), 23 final, Brussels, 27 March.

Commission of the European Communities (CEC 1992b), Communication from the Commission, 'Intellectual Property Rights and Standardization', COM, (92), 445 final, Brussels, 27 October.

Commission of the European Communities (CEC 1990), *Commission Green*

Paper on the Development of European Standardization: Action for Faster Technological Integration in Europe, COM, (90), 456 final, Brussels, 8 October.

Commission of the European Communities (CEC 1988), 'Council Directive on the Limitation into the Air of Certain Pollutants into the Air from Large Combustion Plants', *Official Journal of the European Communities*, L, **336** (1), Brussels, 7 December.

Commission of the European Communities (CEC 1987), *Towards a Dynamic European Economy: Green Paper on the Development of the Common Market for Telecommunications Services and Equipment*, COM, (87), 290 final, Brussels, 30 June.

Commission of the European Communities (CEC 1985), 'Council Resolution', *Official Journal of the European Communities*, C, **136**, 4 June.

Coonley, H. (1956), 'The International Standards Movement', in D. Reck (ed.), *National Standards in a Modern Economy*, New York: Harper & Brothers.

Cordes, J.J. and Tassey, G. (1993), 'Government-Industry Consortia at the National Institute of Standards and Technology: Overcoming the Free-rider Problem', paper presented at the Western Economic Association International, June.

Cowan, R. (1991), 'Technological variety and competition: issues of diffusion and intervention', in OECD (ed.), *Technology and Productivity: the Challenges for Economic Policy*, Paris: OECD.

Cowhey, P.F. (1990), 'The International Telecommunications Regime: the Political Roots of Regimes for High Technology', *International Organization*, **44**, (2), 169–99.

Cowhey, P.F. and Aronson, J.D. (1993), 'A New Trade Order', *Foreign Affairs*, **72**, (1), 183–95.

Crane, R.J. (1979), *The Politics of International Standards: France and the Color TV War*, Norwood, NJ: Ablex Publishing Co.

D'Agostino, R. Jr., and Wilson, R. (1993), 'Asbestos: the Hazard, the Risk, and Public Policy', in K.R. Foster, D.E. Bernstein and P.W. Huber (eds), *Phantom Risk: Scientific Inference and the Law*, Cambridge, MA: MIT Press.

Dankbaar, B. and van Tulder, R. (1991), 'The Influence of Users in Standardization; the Case of MAP', Maastricht Economic Research Institute on Innovation and Technology (MERIT), Working Paper 91–013.

David, P.A. (1994), 'Standardization Policies for Network Technologies: the Flux Between Freedom and Order Revisited', in this volume, Ch. 3.

David, P.A. (1993), 'Path-dependence and Predictability in Dynamic Systems with Local Network Externalities: a Paradigm for Historical Economics,' in D. Foray and C. Freeman (eds), *Technology and the Wealth of Nations*, London: Pinter Publishers.

David, P.A. (1992), 'Why are Institutions the "Carriers of History"? Notes on Path Dependence and the Evolution of Conventions, Organizations and Institutions', Stanford Institute for Theoretical Economics Working Paper, October.

David, P.A. (1991), 'Reputation and Agency in the Historical Emergence of the Institutions of "Open Science"', CEPR Discussion Paper No. 261, Stanford University, CA.

David, P.A. (1987), 'Some New Standards for the Economics of Standardization in the Information Age', in P. Dasgupta and P. Stoneman (eds), *Economic Policy and Technological Performance*, Cambridge: Cambridge University Press.

David, P.A. and Bunn, J.A. (1988), 'The Economics of Gateway Technologies and Network Evolution: Lessons from Electricity Supply History,' *Information Economics and Policy*, **3**, (2), 165–202.

David, P.A. and Foray, D. (1994), 'Markov Random Fields, Percolation Structures and the Economics of EDI Standards Diffusion', in G. Pogorel (ed.), *Global Telecommunication Strategies and Technical Change*, Amsterdam: Elsevier.

David, P.A. and Greenstein, S. (1990), 'The Economics of Compatibility Standards: an Introduction to Recent Research', *Economics of Innovation and New Technology*, **1**, (1/2), 3–41.

David, P.A. and Rothwell, G. (1993), 'Standardization, Diversity, and Learning: a Model for the Nuclear Power Industry', MERIT Research Memorandum 93–005, University of Limburg, Maastricht (forthcoming in *International Journal of Industrial Organization*).

David, P.A. and Steinmueller, W.E. (1993), 'Economics of Compatibility Standards and Competition in Telecommunication Networks', paper prepared for the International Telecommunication Society European Conference, Göteburg, 20–22 June.

David, P.A. and Steinmueller, W.E. (1990), 'The ISDN bandwagon is Coming – Who will be there to Climb Aboard? Quandaries in the Economics of Data Communication networks', *Economics of Innovation and New Technology*, **1**, (1/2), 43–62.

Doern, B. and Phidd, R. (1983), *Canadian Public Policy: Ideas, Strucure and Process*, Toronto: Methuen.

Doern, B. and Purchase B. (eds) (1991), *Canada at Risk; Canadian Public Policy in the 1990s*, Toronto: C.D. Howe Institute.

Doern, B. and Wilson, V.S. (eds) (1974), *Issues in Canadian Public Policy*, Toronto: MacMillan.

Dosi, G. (1982), 'Technological Paradigms and Technological Trajectories', *Research Policy*, **11**, 147–62.

Dosi, G., Pavitt, K. and Soete, L. (1990), 'Technology and Trade: an Overview of the Literature', in G. Dosi, K. Pavitt and L. Soete (eds), *The Economics of Technical Change and International Trade*, Hemel Hempstead: Harvester/Wheatsheaf.

Dosi, G., Teece, D.J. and Winter, S. (1992), 'Toward a Theory of Corporate Coherence: Preliminary Remarks', in G. Dosi, R. Giannetti, and P.A. Toninelli (eds), *Technology and Enterprise in a Historical Perspective*, Oxford: Clarendon Press.

Eicher, L.D. (1990), 'Building Global Consensus for Information Technology Standardization', in J.L Berg and H. Schumny (eds), *An Analysis of the Information Technology Standardization Process*, Amsterdam: North-Holland.

EITO (European Information Technology Organization) (1993), *European Information Technology Observatory 93*, Mainz: Eggebrect-Presse.

ETSI (1992), 'Progress on User Participation in the Standardization Process', Document ITSC3(92)–25, Inter-regional Telecommunications Standards Conference, Tokyo, 5–6 November.

EUREKA Secretariat (1993), *Evaluation of Eureka: Industrial and Economic Effects*, Brussels.

European Community (EC 1985), Council Directive (85/C 136/01) of 7 May on a new approach to technical harmonization and standards.

European Community (EC 1983), Council Directive (83/189/EEC) of 28 March laying down a procedure for the provision of information in the field of technical standards and regulations.

Farrell, J. (1993), 'Choosing the rules for formal standardization', paper presented at NBER Universities Research Conference, Cooperation, Coordination and Collusion among Firms, Cambridge, MA, 14–15 May.

Farrell, J. (1992), 'Some Arguments for Weaker Intellectual Property Protection in Network Industries', Department of Economics, University of California at Berkeley, mimeo.

Farrell, J. and Saloner, G. (1992), 'Converters, Compatibility, and the Control of Interfaces,' *Journal of Industrial Economics*, **40**, (1), 55–83.

Farrell, J. and Saloner, G. (1988), 'Coordination through Committees and Markets', *Rand Journal of Economics*, **19**, Summer, 235–52.

Farrell, J. and Saloner, G (1987), 'Competition, Compatibility and Standards: the Economics of Horses, Penguins and Lemmings', in H.L. Gabel (ed.), *Product Standardisation and Competitive Strategy*, Amsterdam: North-Holland.

Farrell, J. and Saloner, G. (1986), 'Installed Base and Compatibility: Innovation, Product Preannouncements, and Predation', *American Economic Review*, December, **76**, 940–55.

Farrell, J. and Saloner, G. (1985), 'Standardization, Compatibility and Innovation', *Rand Journal of Economics*, **16**, (1), 70–83.

Fisher, L.M. (1993), 'Novell Expected to Open Unix to Standard Setters', *New York Times*, 20 September, D1, D8.

Foray, D. (1994), 'Users, Standards and the Economics of Coalitions and Committees', *Information Economics and Policy*, **6**, (3–4), 269–293.

Foray, D. (1993a), 'The Role of Users in ITS (Information Technology Standardization): Meta-standards, Early Standardization and the Economics of Coalition', paper prepared for the International Telecommunications Society European Conference, Göteburg, 20–22 June.

Foray, D. (1993b), 'Standardisation et Concurrence: des relations ambivalentes', mimeo, Paris: Ecole Centrale.

Foray, D. (1992), 'The Role of Users in IT Standardization', DSTI/ICCP paper, Paris: OECD.

Foray, D. (1990), 'Exploitation des externalités de réseau versus évolution des normes', *Revue d'Economie Industrielle*, **51**, 113–40.

Franke, J.F. (1993), 'Technical Change and Social Regulation: Processes and Institutions', Diskussion Papier 161, Berlin: Technische Universität.

Freeman, C. (1994), 'The Economics of Technical Change: a Critical Survey Article', *Cambridge Journal of Economics*, (forthcoming).

Gabel, H.L. (1987), 'Product Standards and Competitive Strategy: an Analysis of the Principles', mimeo, Fontainebleau: INSEAD.

Genschel, P. and Werle, R. (1992), 'From National Hierarchies to International Standardization: Historical and Modal Changes in the Coordination of Telecommunications', *Max-Planck-Institut für Gesellschaftsforschung Discussion Paper No. 92/1*, Cologne, February.

Glachant, M. (1993), 'Voluntary Agreements in Environmental Policy: Economic Nature and Efficiency', Paris: OECD Report.

Greenstein, S.M. (1993), 'Invisible Hands Versus Invisible Advisors: Coordination Mechanisms in Economic Networks,' Faculty Working Paper 93–0111, Political Economy Series No. 61, Bureau of Economic and Business Research, University of Illinois at Urbana-Champaign, February.

Hahn, R.W. (1993), 'Comparing environmental markets with standards', *Canadian Journal of Economics*, **26**, (2), 346–54.

Hahn, R.W. and Stavins, R.N. (1991), 'Incentive-based Environmental Regulation: a New Era from an Old Idea?', *Ecology Law Quarterly*, **18**, (1), 1–42.

Haigh, N. (1993), *Manual of Environmental Policy: The EC and Britain*, London: Longman.

Haigh, N. (1984), *EC Environmental Policy and Britain: An Essay and a Handbook*, London: Environmental Data Services Ltd.

Haigh, N. and Baldock, D. (1989), *Environmental Policy and 1992*, London: Institute for European Environmental Policy.

Hawkins, R. (1995a), 'Standards-making as Technological Diplomacy: Assessing Objectives and Methodologies in Standards Institutions', in this volume, Ch. 14.

Hawkins, R. (1995b), 'Enhancing the User Role in the Development of Technical Standards for Telecommunication', *Technology Analysis & Strategic Management*, **7**, (1), 21–40.

Hawkins, R. (1993a), 'Public Standards and Private Networks: Some Implications of the "Mobility Imperative"', ENCIP (European Network for Communication and Information Perspectives) Working Paper, Montpellier, March.

Hawkins, R. (1993b), 'Changing Expectations: Voluntary Standards and the Regulation of European Telecommunication', *Communications & Strategies*, No. 11, 3rd Trimestre, September, 53–85.

Hawkins, R. (1992a), 'Standards for Technologies of Communication: Policy Implications of the Dialogue between Technical and Non-technical Factors', Unpublished DPhil Dissertation, SPRU, University of Sussex.

Hawkins, R. (1992b), 'The Doctrine of Regionalism: a New Dimension for International Standardization in Telecommunications', *Telecommunications Policy*, **16**, (4), May/June, 339–53.

Hayden, F.G. (1987), 'Evolution of Time Constructs and their Impact on Socioeconomic Planning', *Journal of Economic Issues*, **21**, (3), 1281–312.

Hellmich, M. (1927), *DIN, 1917–1927*, Berlin: Deutscher Normenausschuss, translated in R.A. Brady (1933), *The Rationalization Movement in German Industry: a Study in the Evolution of Economic Planning*, Berkeley: University of California Press.

Hemenway, D. (1975), *Industry-wide Voluntary Product Standards*, Cambridge: Balinger.

Henderson, R.M. and Clark, K.B. (1990) 'Architectural Innovation: the Reconfiguration of Existing Product Technologies and the Future of Established Firms', *Administrative Science Quarterly*, **35**, 9–30.

Hunter, S. and Waterman, R.W. (1992), 'Determining an Agency's Regulatory Style: How Does the EPA Water Office Enforce the Law?', *Western Political Quarterly*, **2**, June, 403–17.

Information Week (1993), 15 February, 50.

Institut Français de l'Energie (1992), *Reduction of Emissions of Air Pollutants from New and Existing Installations < 50 MW$_{th}$*, Paris, January.

ISO (1986), *100 Year Commemoration for International Standardization*, Geneva: ISO, September.

ISO/IEC (1990), *A Vision for the Future: Standards Needs for Emerging Technologies*, Geneva: ISO/IEC.

ITU (1991), *Tomorrow's ITU: The Challenge of Change*, Report of the High Level Committee to review the structure and functioning of the ITU, Geneva: ITU, April.

Jasanoff, S. (1992), 'What Judges Should Learn from the Sociology of Science', *Jurimetrics Journal*, Spring, **32**, (3), 345–60.

Johansson, S.R. (1993), 'The Brain's Software: the Natural Languages and Poetic Information Processing,' in H. Haken, A. Karlqvist and U. Svedin (eds), *The Machine as Metaphor and Tool*, Berlin: Springer-Verlag.

Johansson, S.R. (1988), 'The Computer Paradigm and the Role of Cultural Information in Social Systems', *Historical Methods, A journal of quantitative and interdisciplinary history*, **21**, 172–88.

Johnson, M. (1993), 'Unix Rivals Unite', *Computerworld*, 22 March, **14**.

Johnson, S.P. and Corcelle, G. (1989), *The Environmental Policy of the European Communities*, London: Graham & Trotman.

Kahaner, D.K. (1991), 'TRON (The Real Time Operating System Nucleus)', *Scientific Information Bulletin*, **16**, (3), 11–19.

Kahneman, D., Knetsch, J.L. and Thaler, R.H. (1991), 'Anomalies: the Endowment Effect, Loss Aversion, and Status Quo Bias', *Journal of Economic Perspectives*, **5**, (1), 193–206.

Kapor, M. (1993), 'Where is the Digital Highway Really Heading? The Case for a Jeffersonian Information Policy', *Wired*, July/August, 53–94.

Katz, M.L. and Shapiro, C. (1986), 'Technology Adoption in the Presence of Network Externalities', *Journal of Political Economy*, **94**, (4), 822–41.

Katz, M.L. and Shapiro, C. (1985), 'Network Externalities, Competition, and Compatibility', *American Economic Review*, June, **75**, 424–40.

Kemp, R. (1992), 'Environmental Policy and Technical Change in Pollution Control: a Critical Assessment of the Existing Theory', mimeo, Limburg: University of Limburg.

Kentridge, C. (1993), 'Environmental Enforcement: Stick vs. Carrot', *Canadian Lawyer*, **17**, (4), 37–42.

Kindleberger, C.P. (1983), 'Standards as Public, Collective and Private Goods', *Kyklos*, **36**, (3), 377–96.

Klein, B., Crawford, R., and Alchian, A. (1978), 'Vertical Integration, Appropriable Rents, and the Competitive Contracting Process,' *Journal of Law and Economics*, **21**, 297–326, reprinted in L. Putterman (1986) (ed.), *The Economic Nature of the Firm*, Cambridge: Cambridge University Press.

Kodama, F. (1991), *Analyzing Japanese High Technologies: The Technoparadigm Shift*, London: Pinter Press.

Krämer, S. (1990), *EEC Treaty and Environmental Protection*, London: Sweet & Maxwell.

Kraus, N., Malmfors, T. and Slovic, P. (1992), 'Intuitive Toxicology: Expert and Lay Judgements of Chemical Risks', *Risk Analysis*, **12**, 215–32.

Krimsky, S. and Golding, D. (1992) (eds), *Social Theories of Risk*, Westport, CT: Praeger.

Laffont, J.-J. and Tirole, J. (1993), *A Theory of Incentives in Procurement and Regulation*, Cambridge: MIT Press.

Lawson, T. (1993), 'Why are so Many Economists so Opposed to Methodology?', paper presented at the European Association of Evolutionary Political Economy Conference, Barcelona, 29–30 October, Cambridge: Faculty of Economics and Politics.

Lehr, W. (1993), 'The Economics of Anticipatory Standard Setting', unpublished paper presented at the Twenty-First Annual Telecommunications Policy Research Conference, 2–4 October, Solomons Island, Maryland.

Lehrer, K. and Wagner, C. (1981), *Rational Consensus in Science and Society*, Dordrecht: D. Reidel.

Leiss, W. (1993), '"Down and Dirty": the Use and Abuse of Public Trust in Risk Communication', paper presented at the Royal Society of Canada symposium on 'Managing Risks to Life and Safety', Ottawa, Ontario, October.

Leiss, W. (1992), *Multi-Stakeholder Negotiation in Environmental Controversies*, Waterloo, Ontario: University of Waterloo, Institute for Risk Research.

Leiss, W. and Chociolko, C. (1993), 'Why is Risk Controversial?', *Risk Management*, **40**, (5), May, 30–35.

Leiss, W. and Chociolko, C. (1994), *Risk and Responsibility*, Montreal: McGill-Queen's University Press.

Leveque, F. (1993a), 'How can Environmental Policy Makers Tackle Industrial Diversity?', in *Environmental Policies and Industrial Competitiveness*, Paris: OECD.

Leveque, F. (1993b), 'The Firm and its Regulatory Environment', mimeo, Paris: CERNA.

Lewontin, R.C. (1982), *Human Diversity*, New York: Scientific American Books.

Link, A.N. and Tassey, G. (1993), 'The Technology Infrastructure of Firms: Investments in Infratechnology', *IEEE Transactions on Engineering Management*, **40**, (3), August, 312–14.

Lundvall, B.-Å. (1992), 'User–Producer Relationships, National Systems of Innovation and Internationalisation', in Lundvall, B.-Å. (ed.), *National Systems of Innovation*, London: Pinter Publishers.

Lundvall, B.-Å. (1988), 'Innovation as an Interactive Process – from User–producer interaction to national systems of innovation', in G. Dosi, C. Freeman, R. Nelson *et al.* (eds), *Technical Change and Economic Theory*, London: Pinter Publishers.

M'Gonigle, R.M., Jamieson, L., McAllister, M.K. and Peterman, R.M. (1995), 'Taking Uncertainty Seriously: from Permissive Regulation to Preventative Design in Environmental Decision-making', *Osgood Law Journal*, **32**, 1, 99–169.

Majone, G. (1978), 'Environmental Standard-setting: Efficiency, Equity and Procedural Problems', Workshop on Environmental Standard-Setting, Laxenburg: IIASA.

Mansell, R. (1993), *The New Telecommunications: a Political Economy of Network Evolution*, London: Sage.

Mansell, R. (1990), 'Rethinking the Telecommunication Infrastructure: the New "Black Box"', *Research Policy*, **19**, (6), 507–15.

Mansell, R. and Hawkins, R. (1992), 'Old Roads and New Signposts: Trade Policy Objectives in Telecommunication Standards', in F. Klaver and P. Slaa (eds), *Telecommunication: New Signposts to Old Roads*, Amsterdam: IOS Press.

Mansfield, E. (1991), 'Academic Research and Industrial Innovation', *Research Policy*, **20**, (2), 1–12.

Margolis, H. (1991), 'Free Riding Versus Cooperation', in R. Zeckhauser (ed.), *Strategy and Choice*, Cambridge, MA: MIT Press.

Metcalfe, J.S. and Boden, M. (1992), 'Evolutionary Epistemology and the Nature of Technology Strategy', in R. Coombs, P. Saviotti and V. Walsh (eds), *Technology Change and Company Strategies: Economic and Sociological Perspectives*, London: Harcourt Brace Jovanovich.

Middleton, R.W. (1980), 'The Gatt Standard Code', *Journal of World Trade Law*, **14**, (3), 201–19.

Monroe, H.K. (1993), 'Mix-and-match Compatibility and Asymmetric Costs', University of Oxford, Unpublished DPhil Dissertation, March.

Morgan, M.G. (1993), 'Risk Analysis and Management', *Scientific American*, **269**, (1), July, 24–30.

Morris, C.R. and Ferguson, C.H. (1993), 'How Architecture Wins the Technology Wars', *Harvard Business Review*, March–April, 86–96.

Nadai, A. (1994), 'The Greening of the EC Agrochemical Market: Regulations and Competition', in *Business Strategy and the Environment*, **3**, (2), Summer, 34–42.

Nadai, A. (1993), 'Les processus communautaires d'évaluation et de signalement au marché des performances environnementales des produits, des sites et des entreprises', mimeo, Paris: CERNA.

Naemura, K. (1992a), 'New Environment of IT Standardization and Changing Roles of Users and Governments', paper presented at the MITI/MPT/OECD workshop 'The Economic Dimension of IT Standards: Users and Governments in the Standardization Process', Tokyo, 17–18 November.

Naemura, K. (1992b), 'Open Horizons and Standardization', paper presented at OECD/ICCP Special Session on IT Policy, Paris, 12–13 October.

Narjes, K.H. (1987), 'Towards a European Telecommunications Community: Implementing the Green Paper', *Telecommunications Policy*, **12**, February, 106–8.

National Academy of Engineering (1993), *Mastering a New Role*, Washington, DC: National Academy Press.

Noam, E. (1992), *Telecommunications in Europe*, Oxford: Oxford University Press.

Nusbaumer, J. (1984), 'The GATT Standards Code in Operation', *Journal of World Trade Law*, **18**, (6), 542–52.

O'Connor, R.M. (1992), 'Lessons from governmental intervention in OSI', paper presented at the MITI/MPT/OECD workshop, 'The Economic Dimension of IT Standards: Users and Governments in the Standardization Process', Tokyo, 17–18 November.

O'Connor, R.M. (1988), 'The Economic Impact of IT Standards,' in N.E. Malgardis and T.J. Williams (eds), *Standards in Information Technology and Industrial Control*, Amsterdam: Elsevier.

OECD (1993a), *The 1992/93 Communications Outlook*, Paris: OECD.

OECD (1993b), *Environmental Policies and Industrial Competitiveness*, Paris: OECD.

OECD (1991), *Information Technology Standards: The Economic Dimension*, DSTI/ICCP Report No. 25, Paris: OECD.

Partridge, P.H. (1971), *Consent and Consensus*, London: Pall Mall.

Pelkmans, J. (1987), 'The New Approach to Technical Harmonization and Standardization,' *Journal of Common Market Studies*, **25**, (3), 249–69.

Pelkmans, J. and Beuter, R. (1986), 'Standardization and Competititiveness: Private and Public Strategies in the EC Colour TV Industry', paper for the INSEAD Symposium, in H.L. Gabel (ed.), *Product Standardization as a Tool of Competitive Strategy*, Amsterdam: Elsevier Science Publishers.

Peltzman, S. (1976), 'Toward a More General Theory of Regulation', *Journal of Law and Economics*, **19**, (2), 211–40.

Penrose, E.T. (1959), *The Theory of the Growth of the Firm*, New York: Wiley.

Pipe, R. (1993), 'Trade of Telecommunications Services: Implications of a GATT Uruguay Round Agreement for ITU and Member States', report prepared for Strategic Planning Unit, International Telecommunication Unit, Geneva, May.

Prins, C. and Schiessl, M. (1993), 'The New European Telecommunications Standards Institute Policy: Conflicts Between Standardisation and Intellectual Property Rights', *European Intellectual Property Review*, **15**, (8), 263–66.

Pritzker, D.M. and Dalton, D.S. (1990) (eds), *Negotiated Rulemaking Sourcebook*, Washington DC: Administrative Conference of the United States.

Puffert, D.J. (1991), 'The Economics of Spatial Network Externalities and the Dynamics of Railway Gauge Standardization', unpublished PhD Dissertation, Stanford University.

Quintas, P. (1994), 'Paths of innovation in Software and Systems Develop-
ment Practice', in R. Mansell (ed.), *The Management of Information and
Communication Technologies: Emerging Patterns of Control*, London:
ASLIB.

Rankine, L.J. (1992), 'The Role of Users in IT Standardization', DSTI/ICCP
paper, Paris: OECD.

Rankine, L.J. (1990), 'Information Technology Standards – can the challenges
be met?', in J.L. Berg and H. Schumny (eds), *An Analysis of the Information
Technology Standardization Process*, Amsterdam: North-Holland.

Reddy, M.N. (1987), 'Technology, Standards, and Markets: a Market Institutionalization Perspective', in H.L. Gabel (ed.), *Product Standardization
and Competitive Strategy*, Amsterdam: Elsevier Science Publishers.

Renn, O. (1992), 'Risk Communication: Toward a Rational Discourse with
the Public', *Journal of Hazardous Materials*, **29**, 465–519.

Reynolds, P. (1990), 'Buyers, Sellers & Standards – What Should Both Sides
Do Differently?', in J.L. Berg and H. Schumny (eds), *An Analysis of the
Information Technology Standardization Process*, Amsterdam: North-
Holland.

Rosenberg, N. (1990), 'Why do Firms do Basic Research (with their Own
Money)?', *Research Policy*, **19**, (2), 165–74.

Rosenberg, N. (1982), *Inside the Black Box: Technology and Economics*,
Cambridge: Cambridge University Press.

Ruby, P. and Salter, A. (1993), 'Strategic Alliances and Standardization',
report prepared for the Standards Program Office, Department of Communications, Ottawa.

Ruggie, J. G. (1993), *Multilateralism Matters: Theory and Praxis of Institutional Form*, New York: Columbia University Press.

Saloner, G. (1990), 'Economic Issues in Computer Interface Standardization', *Economics of Innovation and New Technology*, **1**, (1/2), 135–56.

Salter, L. (1993–94a), 'Have We Reached the Information Age yet?', in
'Have We Reached the Information Age yet? The Political Economy of
Information Standards', *International Journal of Political Economy*, **23**,
(4), special issue, 3–25.

Salter, L. (1993–94b), 'The Housework of Capitalism', in 'Have We Reached
the Information Age yet? The Political Economy of Information Standards', *International Journal of Political Economy*, **23**, (4), special issue,
105–31.

Salter, L. (1988), *Mandated Science: Science and Scientists in the Making of
Standards*, Boston: Kluwer Academic Publishers.

Samuelson, P.A. (1954), 'The Pure Theory of Public Expenditure', *Review of
Economics and Statistics*, **36**, November, 387–89.

Schelling, T. (1978), *Micromotives and Macrobehavior*, New York: Norton.

Schmidt, S. (1993), 'Coordinating Complementarities: the Case of Institutionalised Standardisation', paper prepared for the Research Seminar on 'Institutional Change and Network Evolution', Stockholm, 16–18 June.

Schmidt, S. (1992), 'Negotiating Technical Change through Standards: Technical Coordination in Markets and Committees', Cologne: Max-Planck-Institut.

Schmidt, S. and Werle, R. (1992), 'The Development of Compatibility Standards in Telecommunications: Conceptual Framework and Theoretical Perspectives', in M. Dierkes and U. Hoffman (eds), *New Technology at the Outset*, Frankfurt/New York: Campus Verlag.

Schmidt, S. and Werle, R. (1993), 'Technical Controversy in International Standardization', Max-Planck-Institut für Gesellschaftsforschung Discussion Paper No. 93/5, Cologne, March.

Schumpeter, J.A. (1954), *A History of Economic Analysis*, New York: Oxford University Press.

Scitovsky, T. (1976), *The Joyless Economy: an Inquiry in Human Satisfaction and Consumer Dissatisfaction*, Oxford: Oxford University Press.

Shankar, N. (1992), 'Setting a New Standard', *Communications International*, 19 December, 12–13.

Shepherd, W.G. (1984), "Contestability' vs. competition,' *American Economic Review*, **74**, (September), 572–85.

Shurmer, M. (1993), 'International Telecommunications Standardization: an Overview of the Current Institutional Framework and Evaluation of Standards Policy', paper prepared for the International Telecommunications Society European Conference, Göteburg, 20–22 June.

Sinclair, B. (1969), 'At the Turn of a Screw: William Sellers, the Franklin Institute, and a Standard American Thread', *Technology and Culture*, **10**, (1), 20–34.

Skea, J. (1993), 'Regulatory Dynamics: Business and the Genesis of the European Community Carbon Tax Proposal', 2nd International Conference of the Greening of Industry Network, Designing the Sustainable Enterprise, Boston, 14–16 November.

Spradley, J.P. (1979), *The Ethnographic Interview*, New York: Holt Rinehart & Winston.

Spring, M.B. (1991), 'Information Technology Standards', in M. Williams (ed.), *Annual Review of Information Science and Technology*, **26**, 79–111.

Spulber, D.F. (1989), *Regulation and Markets*, Cambridge: MIT Press.

Stavins, R.N. and Whitehead, B.W. (1992), 'Dealing with Pollution: Market-based Incentives for Environmental Protection', *Environment*, **34**, (7), 6–1, 29–45.

Steinmueller, W.E. (1995), 'The Political Economy of Data Communication Standards', in this volume, Ch. 16.

Steinmueller, W.E. (1994), 'Centralization vs. Decentralization of Data Communications Standards', in G. Pogorel (ed.), *Global Telecommunications Strategies and Technological Changes*, Amsterdam: North-Holland.

Steinmueller, W.E. (1992), 'The Economics of Production and Distribution of User-specific Information via Digital Networks', in C. Antonelli (ed.), *The Economics of Information Networks*, Amsterdam: North-Holland.

Stigler, G.J. (1971), 'The Theory of Economic Regulation', *Bell Journal of Economic and Management Science*, **2**, 3–21.

Strawbridge, G.M. (1991), 'The New Approach to Technical Harmonization and Standards', London: British Standards Institution, mimeo.

Swann, P. (1992), 'Standards in ICT: Consensus, Institutions and Markets', in G. Locksley (ed.), *The Single European Market and the Information and Communications Technologies*, London: Belhaven Press.

Swann, P. (1991), 'User's Needs for Standards: How Can We Ensure that User's Votes are Counted?', in B. Meek (ed.), *User's Needs in Standards*, London: Butterworth.

T1 (1992), 'Promoting user involvement', Document ITSC3(92)–06, Interregional Telecommunications Standards Conference, Tokyo, 5–6 November.

Tamarin, C. (1988), 'Telecommunications Technology Applications and Standards: a New Role for the User', *Telecommunications Policy*, December, 323–31.

Tang, P. and Mansell, R. (1993), 'Telecommunication, Multinational Enterprises and Globalization: Implications for Future Network Development', report prepared for Rank Xerox Cambridge EuroParc, February.

Task Force in the Environment and the Internal Market (1990), *'1992': The Environmental Dimension*, Bonn: Economica Verlag.

Tassey, G. (1992), *Technology Infrastructure and Competitive Position*, Amsterdam: Kluwer Academic Publishers.

Teece, D. J. (1987), 'Profiting from Technological Innovation: Implications for Integration, Collaboration, Licensing, and Public Policy,' in D.J. Teece (ed.), *The Competitive Challenge: Strategies for industrial innovation and renewal*, Cambridge: Ballinger.

Thompson, G.V. (1954), 'Intercompany Technical Standardization in the Early American Automobile Industry', *Journal of Economic History*, **24**, (1), 1–20.

TTC (1992), 'User Participation in TTC standardization activities', Document ITSC3(92)–19, Interregional Telecommunications Standards Conference, Tokyo, 5–6 November.

Ungerer, H. (1990), *Telecommunications in Europe: Free Choice for the User in Europe's 1992 Market*, Brussels: Commission of the European Communities, The European Perspectives Series.

US Congress, Office of Technology Assessment (1992), *Global Standards: Building Blocks for the Future*, TCT–512, Washington, DC: Government Printing Office, March.

Utterback, J.M. and Suarez, F.F. (1993), 'Innovation, Competition and Industry Structure', *Research Policy*, **22**, (1), 61–81.

Van Vliet, M. (1992), 'Communicative Governance in Complex Networks: Options for Environmental Regulation of Business', mimeo, Rotterdam: Erasmus Universiteit.

Verman, L.C. (1973), *Standardization: a New Discipline*, Hamden: Archon Books.

Victor, P.W. (1979), 'Economics and the Challenge of Environmental Issues', in W. Leiss, (ed.), *Ecology versus Politics in Canada*, Toronto: University of Toronto Press.

von Hippel, E. (1988), *The Sources of Innovation*, Oxford: Oxford University Press.

von Hippel, E. (1985), 'Learning from lead users', in R.D. Buzzell (ed.), *Marketing in an Electronic Age*, Cambridge, MA: Harvard Business School Press.

Wallenstein, G. (1990), *Setting Global Telecommunications Standards*, Norwood, MA: Artech House.

Weiss, M.B.H. and Toyofuku, R.T. (1993), 'Free-ridership in the Standards-setting Process: the Case of 10BaseT', Carnegie Mellon University, mimeo, 1993.

Weiss, M.B.H. and Spring, M. (1992), 'Selected Intellectual Property Issues in Standardization', paper for the 12th Annual Telecommunications Policy Research Conference, Solomons MD, September.

Weiss, M.B.H. and Sirbu, M. (1990), 'Technological Choice in Voluntary Standards Committees: an Empirical Analysis', *Economics of Innovation and New Technology*, **1**, (1/2), 111–33.

Weitzman, M.L. (1974), 'Prices vs. quantities', *Review of Economic Studies*, **4**, (41), October.

Whitworth, J. (1882), 'Papers on Mechanical Subjects, Part I: True Planes, Screw Threads and Standard Measures', London: E.F. and N. Spon.

Williams, R. (1976), *Keywords*, London: Fontana, Croom Helm.

Williamson, O.E. (1971), 'The Vertical Integration of Production: Market Failure Considerations,' *American Economic Review*, **61**, May, 112–23.

Willinger, M. (1993), 'L'internalisation de l'environnement dans les organisations industrielles: une mise en perspective méthodologique', 8th Mediterranean Summer School of Industrial Economics, Cargèse: 20–25 September.

Winter, S.G. (1987), 'Knowledge and Competence as Strategic Assets,' in

D.J. Teece (ed.), *The Competitive Challenge: Strategies for Industrial Innovation and Renewal*, Cambridge: Ballinger.

Yamada, K. (1993), 'Apple is Developing Software that lets Workstations run Macintosh Programs', *Wall Street Journal*, May 11, B6.

Young, A.A. (1928), 'Increasing Returns and Economic Progress, *Economic Journal*, **38**, 527–42.

Zartman, I.W. (1977), 'Negotiation as a Joint Decision-making Process', in I.W. Zartman (ed.), *The Negotiation Process: Theories and Applications*, Beverly Hills: Sage.

Author index

Subject index